T0211859

Springer Theses

Recognizing Outstanding Ph.D. Research

More information about this series at http://www.springer.com/series/8790

Aims and Scope

The series "Springer Theses" brings together a selection of the very best Ph.D. theses from around the world and across the physical sciences. Nominated and endorsed by two recognized specialists, each published volume has been selected for its scientific excellence and the high impact of its contents for the pertinent field of research. For greater accessibility to non-specialists, the published versions include an extended introduction, as well as a foreword by the student's supervisor explaining the special relevance of the work for the field. As a whole, the series will provide a valuable resource both for newcomers to the research fields described, and for other scientists seeking detailed background information on special questions. Finally, it provides an accredited documentation of the valuable contributions made by today's younger generation of scientists.

Theses are accepted into the series by invited nomination only and must fulfill all of the following criteria

- They must be written in good English.
- The topic should fall within the confines of Chemistry, Physics, Earth Sciences, Engineering and related interdisciplinary fields such as Materials, Nanoscience, Chemical Engineering, Complex Systems and Biophysics.
- The work reported in the thesis must represent a significant scientific advance.
- If the thesis includes previously published material, permission to reproduce this must be gained from the respective copyright holder.
- They must have been examined and passed during the 12 months prior to nomination.
- Each thesis should include a foreword by the supervisor outlining the significance of its content.
- The theses should have a clearly defined structure including an introduction accessible to scientists not expert in that particular field.

Kiyotaka Akabori

Structure Determination of HIV-1 Tat/Fluid Phase Membranes and DMPC Ripple Phase Using X-Ray Scattering

 Springer

Author
Kiyotaka Akabori
Carnegie Mellon University
Pittsburgh, PA, USA

Supervisor
Dr. Stephanie Tristram-Nagle
Department of Physics
Carnegie Mellon University
Pittsburgh, PA, USA

ISSN 2190-5053   ISSN 2190-5061 (electronic)
Springer Theses
ISBN 978-3-319-37047-7   ISBN 978-3-319-22210-3 (eBook)
DOI 10.1007/978-3-319-22210-3

Supervisor's Foreword

In physics, a graduate student has many, diverse fields of study from which to choose for the Ph.D. thesis. These fields range from astrophysics and cosmology, to computational physics, to high-energy experiment or theory, to quark interaction experiment or theory, to condensed matter theory or experiment. Biological physics is another, more recent choice in many physics departments. Biological physics has several subfields, such as biomembrane structure, dynamics and organization (both experiment and theory), folding and unfolding of single protein, DNA and RNA molecules, collective vibrational modes in biomacromolecules, transport and rheology in biopolymer gels, and interaction and structure of supramolecular assemblies. The two focuses of the current thesis are both in the area of biomembranes. Why study biomembranes? Membranes play a central role in both the structure and function of all cells, plant and animal. Membranes not only define compartments, they also determine the nature of all communication between the inside and the outside. In addition, most of the fundamental biochemical functions in cells involve membranes at some point, including such diverse processes as prokaryotic DNA replication, protein biosynthesis, protein secretion, bioenergetics, and hormonal responses. Although biomembranes have been studied since the beginning of the twentieth century, there are many mysteries that remain: precise structure and physical properties of single component lipid membranes, the reason for the large diversity of lipids in biomembranes, and the details of complex interactions or proteins and peptides with biomembranes. The techniques developed by physicists are uniquely suited to probe these underlying questions. Two interesting reference books in this field are *Biomembranes, Molecular Structure and Function*, Robert B. Gennis, Springer, 1989, New York, and *Lipids & Membrane Biophysics*, Faraday Discussions, Vol. 161, Royal Society of Chemistry, 2013, Cambridge, England.

The first focus of the current thesis is Tat, the transactivator of transcription, an important protein for HIV-1 infection. Tat acts by enhancing the readout of HIV-1 RNA, through molecular interactions with the HIV-1 DNA. In order to carry out its function, Tat translocates across the T-cell's nuclear membrane, by relying on a highly positively charged, basic region of only 11 amino acids that can translocate through membranes without requiring energy. It is counterintuitive that a highly charged molecule could not only translocate itself through the low dielectric medium of the hydrocarbon interior of lipid membranes but that it can also be engineered to pull larger, uncharged molecules through membranes as well. Many investigations have attempted to probe Tat's structure in membranes in order to understand its translocation through them. In this thesis, high intensity synchrotron X-rays were used to probe the molecular details of the structural and elastic interactions of Tat with membranes composed of several types of lipids. These experiments were directly compared with atomistic molecular dynamics simulations of Tat interacting with membranes. This comparison delineated the precise location of Tat in biomembranes and its effect on the membranes.

The second focus of the current thesis is the enigmatic ripple phase. While it is not of physiological importance, since it is primarily gel-like in character and since it only forms in single component membranes, it has been the subject of many theoretical and experimental physics papers as an example of a periodically modulated phase. Despite many systematic studies over the past three decades, molecular details of the structure were still lacking, which impeded theoretical understanding of its origin. In this thesis, Dr. Akabori used synchrotron X-rays to probe the ripple phase in the lipid DMPC, oriented onto a silicon wafer and hydrated through the vapor in a hydration chamber. These techniques allowed him to quantitate the degree and direction of chain tilt in both the major and minor arms of the corrugated, sawtooth pattern. A new model of the ripple phase emerged which will serve to motivate theoreticians to supply a driving force.

Pittsburgh, PA, USA   Stephanie Tristram-Nagle
2015/5/29

Abstract

This thesis employs X-ray scattering to study the structure of two different stacked lipid membrane systems. The first part reports the effect on lipid bilayers of the Tat peptide $Y_{47}GRKKRRQRRR_{57}$ from the HIV-1 virus transactivator of transcription (Tat) protein. Synergistic use of low-angle X-ray scattering (LAXS) and atomistic molecular dynamics (MD) simulations indicated Tat peptide binding to neutral dioleoylphosphatidylcholine (DOPC) headgroups. This binding induced the nearby lipid phosphate groups to move 3 Å closer to the bilayer center. Many of the Tat arginines were as close to the bilayer center as the locally thinned lipid phosphate groups. Analysis of LAXS from DOPC, DOPC/dioleoylphosphatidylethanolamine (DOPE), DOPC/dioleoylphosphatidylserine (DOPS), and a mimic of the nuclear membrane indicated that the Tat peptide decreased the bilayer bending modulus K_c and increased the area per lipid, possibly facilitating Tat membrane translocation. Although a mechanism for translocation remains elusive, this study suggests that Tat translocates from the headgroup region.

The second study presents the structure of the asymmetric ripple phase formed by dimyristoylphosphatidylcholine. We determined the most detailed ripple phase structure by combining synchrotron LAXS and wide-angle X-ray scattering (WAXS) from highly aligned multilamellar samples. We derived three intensity corrections to calculate the X-ray form factors from the 52 measured reflections. The LAXS analysis provided a high-resolution two-dimensional electron density map. The ripple major arm was demonstrated to be consistent with the gel phase, and the major and minor arm structures were clearly different, supporting the coexistence of different molecular organizations. The minor arm electron density profile was qualitatively consistent with interdigitated chain packing previously proposed by MD simulations. Analysis of high-resolution near grazing incidence WAXS showed that major arm hydrocarbon chains were tilted parallel to the ripple plane by 18° with respect to the bilayer local normal, toward the next nearest neighbor similarly to the gel $L_{\beta F}$ rather than the $L_{\beta I}$ phase. By measuring the Bragg rod lengths in transmission WAXS, we determined that major arm chains in opposing leaflets were coupled. The LAXS and WAXS results together indicated that chains in the major arm were shorter by 1.3 Å compared to the gel phase, suggesting a gauche-trans-gauche kink in the ripple major arm. In contrast to the LAXS analysis, the measured nGIWAXS was consistent with disordered chains in the minor arm similarly to the fluid L_α phase.

Acknowledgments

I would like to thank my advisors Profs. John F. Nagle and Stephanie Tristram-Nagle for giving me an opportunity to work in the field of biophysics, for guiding my research, and for supporting me as a research assistant.

I especially thank Michael S. Jablin for a number of exciting discussions and projects we shared. Despite our differences, he always respected my opinions and helped me shape my ideas.

I thank Profs. Angel E. García, Robert M. Suter, and Frank Heinrich for serving on my Ph.D. committee.

Thanks to Dr. Kun Huang for providing us with many simulation results.

I thank Prof. Christian D. Santangelo for giving me a chance to work on a theoretical problem.

I thank my parents for allowing me to study at UCSB and financially supporting me. Many thanks to my wife for the countless moments of joy and laughter, supporting my Ph.D. study with great meals every day, and being a wonderful mother. And to Shyn, I am so proud to be your father.

Contents

Notations

Acronyms

Chapter 1
Introduction

Abstract This thesis has two focuses, both in the area of biomembranes. One focus is on the interaction of a biomedically important Tat peptide with membranes. The other is on a fundamental problem regarding the enigmatic structure of a pure lipid bilayer. Section 1.1 introduces lipid molecules that constitute biomembranes and three thermodynamic phases displayed by lipids pertinent to this thesis. The Tat peptide and its biomedical importance are introduced in Sect 1.2, followed by a brief overview of the ripple phase in Sect. 1.3.

1.1 Lipid Bilayers

Membranes define the boundary between living cells and their surrounding environment, and from this position help regulate intercellular transport. The lipid bilayer is the structural backbone of biomembranes. Lipid bilayers are a self-assembly of lipids, which are amphiphilic molecules that consist of a hydrophilic headgroup and hydrophobic chains (Fig. 1.1).

In water, lipids self-assemble into lipid bilayers to shield their hydrophobic chains, and display a wide variety of thermodynamic phases as a function of temperature and hydration. Figure 1.2 shows a phase diagram of dimyristoylphosphatidylcholine (DMPC). Phosphatidylcholine (PC) lipids constitute a substantial fraction of cell membranes and have been studied for many decades [1]. At full hydration (100 % relative humidity), a lamellar phase coexists with excess water. In the high temperature, fluid L_α phase, the hydrocarbon chains are conformationally disordered, and intra-membrane molecular correlations are liquid-like [2] (Fig. 1.3). The disorder of fluid phase membranes allows proteins to interact with cell membranes in various ways, rendering biological systems highly complex. This phase is usually considered the most biologically relevant.

In the low temperature, gel $L_{\beta'}$ phase, hydrocarbon chains are extended in essentially all-trans configuration and tilted with respect to the membrane normal [4] and are organized in either a hexagonal or orthorhombic lattice (Fig. 1.3). The $L_{\beta'}$ phase is further categorized into three phases according to the chain tilt direction [3, 5, 6]. In the $L_{\beta I}$ phase, chains are tilted toward the nearest neighbor as shown in

© Springer International Publishing Switzerland 2015

K. Akabori, *Structure Determination of HIV-1 Tat/Fluid Phase Membranes and DMPC Ripple Phase Using X-Ray Scattering*, Springer Theses, DOI 10.1007/978-3-319-22210-3_1

Fig. 1.1 Schematic
representation of a lipid
molecule

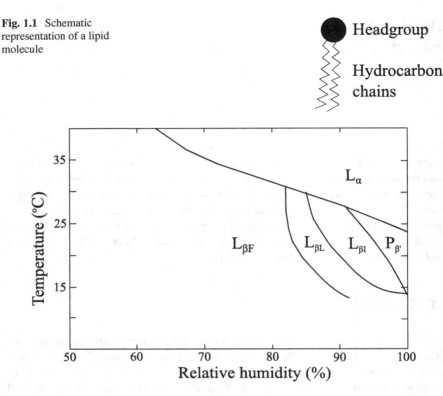

Fig. 1.2 Experimental phase diagram of DMPC [3]. $L_{\beta I}$, $L_{\beta L}$, and $L_{\beta F}$ belong to the gel $L_{\beta'}$ phase. $P_{\beta'}$ is the ripple phase, and L_α is the fluid phase

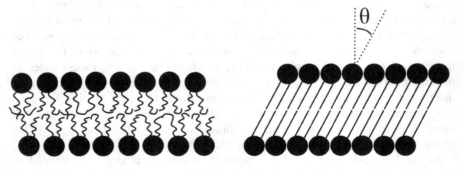

Fig. 1.3 Schematics of the structure of fluid L_α phase (*left*) and gel $L_{\beta'}$ phase (*right*). *Black solid circles* are lipid headgroups and *solid lines* are lipid chains. θ is the chain tilt angle

Fig. 1.4, and in the $L_{\beta F}$ phase, chains are tilted toward the next nearest neighbor. In the $L_{\beta L}$ phase, chains are tilted toward an intermediate direction between nearest and next nearest neighbors.

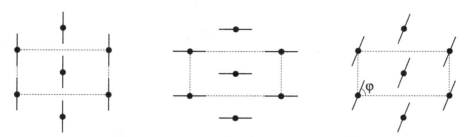

Fig. 1.4 Gel $L_{\beta'}$ phase chains projected onto the bilayer plane showing the chain tilt direction in $L_{\beta I}$ (*left*), $L_{\beta F}$ (*middle*), and $L_{\beta L}$ (*right*) phases. *Black dots* are orthorhombic lattice points. Unit cells are shown in *dashed lines*. Chains are drawn as *solid lines*. Chains are tilted toward the nearest neighbor in $L_{\beta I}$ phase with $\phi = \pi/2$. In the $L_{\beta F}$ phase, the chains are tilted toward the next nearest neighbor ($\phi = 0$). In the $L_{\beta L}$ phase, $0 \le \phi \le \pi/2$

There are various kinds of lipids. They can be categorized in terms of head-group, chain saturation, and chain length. The most studied headgroup is phos-phatidylcholine (PC), consisting of phosphate and choline molecular groups. Lipid hydrocarbon chains can have one or more double bonds. Lipids with no double bonds in the chains are called saturated lipids, such as DMPC (see Fig. 1.5) and dipalmitoylphosphatidylcholine (DPPC). Lipids with one double bond are called mono-unsaturated lipids, such as dioleoylphosphatidylcholine (DOPC) shown in Fig. 1.5. Unsaturation leads to chain packing frustration and lowers the melting temperature. For example, at full hydration DOPC forms a L_α fluid phase at room temperature while DPPC is in a $L_{\beta'}$ gel phase. In mammalian cells, most lipids have at least one unsaturated chain. Membrane curvature has interested many physicists. Phosphatidylethanolamine (PE) is a small headgroup, and packing of PE lipids leads to spontaneous membrane curvature. The chemical structure of dioleoylphos-phatidylechanolamine (DOPE) is shown in Fig. 1.5. Many proteins have been found to sense/induce membrane curvature, making PE lipids especially attractive for those studies [7]. Another class of headgroup is anionic, such as phosphatidylserine (PS) and phosphatidylglycerol (PG). In cells, electrostatic interactions significantly influence biological processes and naturally occurring anionic lipids have been the focus of many studies [8].

1.2 Tat Peptide

The transactivator of translation (Tat), an important protein for HIV-1 infection, is produced by the HIV-1 Tat regulatory gene. After synthesis on the HIV-1 RNA, Tat protein enters a cell's nucleus where it is a transcriptional transactivator for the long terminal repeat promoter which acts by binding to the Tar RNA element [9] (Fig. 1.6). More recently, it was discovered that Tat participates in RNA initiation by stimulating the transcription complex [10]. Both of these roles activate the HIV

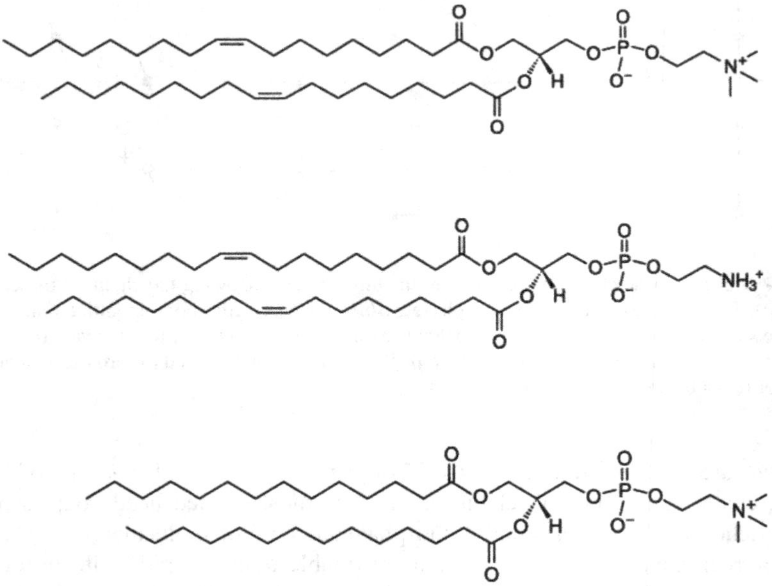

Fig. 1.5 Lipid structures of DOPC (*top*), DOPE (*middle*), and DMPC (*bottom*) (Images are from Avanti Polar Lipids (http://avantilipids.com/))

virus and increase viral loads. One focus of current AIDS research is to eradicate reservoirs of infected memory T-cells that contain dormant HIV; Tat can awaken latent provirus [11]. An understanding of Tat transport could lead to new or more effective HIV treatments. Tat could be prevented from reaching the nuclear genome. Tat could awaken dormant virus so that it can be targeted by standard treatments that only work on active virus [11].

Of the 86 amino acids in Tat, the highly basic (Y_{47}GRKKRRQRRR$_{57}$) sequence called Tat peptide is essential to transport a Tat protein through the nuclear membrane [12, 13]. Mutations within this region yield a Tat protein that does not penetrate the cell nucleus. Tat peptide membrane translocation efficiency has made it a model for peptide-aided protein and drug delivery [14]. The mechanism of Tat-peptide translocation of proteins, DNA, RNA, and drugs across the membrane is of considerable interest since it is known that desolvating and moving charged groups across membranes can be energetically prohibitive [15]. It has been suggested by MD (molecular dynamics) simulations that Tat peptide first binds rather more deeply in the membrane, below the phosphates, than would be anticipated for such a highly charged peptide. From that position, Tat may electrostatically attract the phosphates in the distal monolayer leading to the formation of a transient water-filled pore through which proteins and drugs diffuse [16]. We studied the transverse location of Tat within model lipid membranes by X-ray scattering combined with MD simulations. This study is described in Chap. 2.

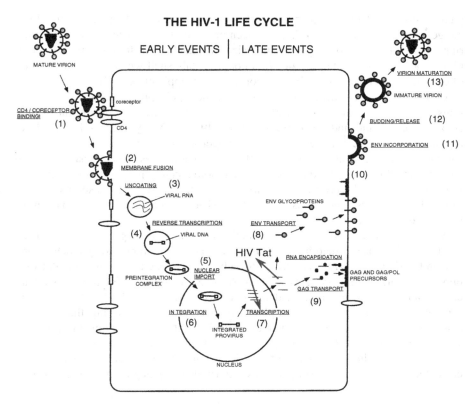

THE HIV-1 LIFE CYCLE

EARLY EVENTS | LATE EVENTS

Fig. 1.6 HIV-1 life cycle (Adapted from Ref. [17] (HIV Tat added with large arrows))

1.3 $P_{\beta'}$ Ripple Phase

For some lipids, a height modulated phase where bilayers are no longer flat exists between the fluid and gel phases (Fig. 1.2). This phase was termed $P_{\beta'}$ and is commonly called the ripple phase [4]. The $P_{\beta'}$–$L_{\beta'}$ transition is often called the pre-transition [18] or lower transition [1]. The ripple phase has fascinated many researchers in condensed matter physics and physical chemistry as an example of periodically modulated phases, with many theoretical papers attempting to explain the height modulation [19–28] and a few simulation papers investigating molecular organization [29–32]. Despite many systematic studies over the past three decades [4, 18, 33–39], molecular details of the structure are still lacking, which impedes theoretical understanding of its origin.

Studies of the ripple phase are normally done on multilamellar systems, but some works have reported the existence of the ripple phase in unilamellar vesicles [40, 41]. Most studies have been performed on PCs [1]. PCs have a fairly bulky headgroup, creating a size mismatch with its acyl chains, especially below the main phase transition. This is believed to be the reason why the acyl chains are tilted in the gel phase of PCs [1, 42, 43]. It has been proposed that the driving force for the ripple formation is also coupled to this size mismatch, with headgroup hydration playing an important role [22, 44]. It is not yet established whether the chains are tilted with respect to the local bilayer normal in the ripple phase [37].

Generally, it is assumed that the lipids in the ripple phase are mainly in all-trans configuration, as in the gel phase [39]. However, many studies point to the coexistence of fluid and gel regions [37, 45–49]. An X-ray structural study has reported that the ripples are composed of a longer sawtooth arm with characteristics similar to a gel phase and a shorter arm that is thinner and less densely packed [37], more compatible with a fluid phase or with a more recently proposed interdigitated L_I phase. Changes of bilayer packing along the ripple direction were also reported in molecular dynamics simulations [29]. Yet, coexistence of different bilayer packings has not been established.

We studied the electron density distribution and chain packing of the asymmetric DMPC ripple phase formed by an oriented multilamellar sample using synchrotron low and wide angle X-ray scattering. An advantage of studying multilamellar (as opposed to unilamellar) systems with X-rays is out-of-plane diffraction peaks that can be analyzed to determine a detailed bilayer structure [37, 39]. An oriented sample also yields anisotropic in-plane chain correlation scattering that can be analyzed to elucidate the molecular organization within the rippling bilayers, as was successfully done for the gel phase [3–5]. Our aim was to study whether all-trans chains are tilted with respect to the local bilayer normal and to elucidate the coexistence of different bilayer packings. My ripple study is presented in Chap. 3.

Bibliography

1. J.F. Nagle, S. Tristram-Nagle, Structure of lipid bilayers. Biochim. Biophys. Acta (BBA) – Rev. Biomembr. **1469**(3), 159–195 (2000)
2. P.F. Fahey, W.W. Webb, Lateral diffusion in phospholipid bilayer membranes and multilamellar liquid crystals. Biochemistry **17**(15), 3046–3053 (1978)
3. G.S. Smith, E.B. Sirota, C.R. Safinya, N.A. Clark, Structure of the $L_{\beta'}$ phases in a hydrated phosphatidylcholine multimembrane. Phys. Rev. Lett. **60**, 813–816 (1988)
4. A. Tardieu, V. Luzzati, F. Reman, Structure and polymorphism of the hydrocarbon chains of lipids: a study of lecithin-water phases. J. Mol. Biol. **75**(4), 711–733 (1973)
5. S. Tristram-Nagle, R. Zhang, R.M. Suter, C.R. Worthington, W.J. Sun, J.F. Nagle, Measurement of chain tilt angle in fully hydrated bilayers of gel phase lecithins. Biophys. J. **64**(4), 1097–1109 (1993)
6. S. Tristram-Nagle, Y. Liu, J. Legleiter, J.F. Nagle, Structure of gel phase DMPC determined by X-ray diffraction. Biophys. J. **83**(6), 3324–3335 (2002)
7. L.V. Chernomordik, M.M. Kozlov, Protein-lipid interplay in fusion and fission of biological membranes. Annu. Rev. Biochem. **72**(1), 175–207 (2003)
8. W. Dowhan, Molecular basis for membrane phospholipid diversity: why are there so many lipids? Annu. Rev. Biochem. **66**(1), 199–232 (1997)
9. Y.N. Vaishnav, F. Wong-Staal, The biochemistry of AIDS. Annu. Rev. Biochem. **60**(1), 577–630 (1991)

10. T. Raha, S.G. Cheng, M.R. Green, HIV-1 Tat stimulates transcription complex assembly through recruitment of TBP in the absence of TAFs. PLoS Biol. **3**(2), e44 (2005)

11. D. Macías, R. Oya, L. Saniger, F. Martín, F. Luque, A lentiviral vector that activates latent human immunodeficiency virus-1 proviruses by the overexpression of Tat and that kills the infected cells. Hum. Gene Therapy **20**(11), 1259–1268 (2009)

12. S. Ruben, A. Perkins, R. Purcell, K. Joung, R. Sia, R. Burghoff, W. Haseltine, C. Rosen, Structural and functional characterization of human immunodeficiency virus Tat protein. J. Virol. **63**(1), 1–8 (1989)

13. J. Hauber, A. Perkins, E.P. Heimer, B.R. Cullen, Trans-activation of human immunodeficiency virus gene expression is mediated by nuclear events. Proc. Natl. Acad. Sci. **84**(18), 6364–6368 (1987)

14. S. Futaki, T. Suzuki, W. Ohashi, T. Yagami, S. Tanaka, K. Ueda, Y. Sugiura, Arginine-rich peptides an abundant source of membrane-permeable peptides having potential as carriers for intracellular protein delivery. J. Biol. Chem. **276**(8), 5836–5840 (2001)

15. M. Grabe, H. Lecar, Y.N. Jan, L.Y. Jan, A quantitative assessment of models for voltage-dependent gating of ion channels. Proc. Natl. Acad. Sci. **101**(51), 17640–17645 (2004)

16. H.D. Herce, A.E. Garcia, Molecular dynamics simulations suggest a mechanism for translocation of the HIV-1 Tat peptide across lipid membranes. Proc. Natl. Acad. Sci. **104**(52), 20805–20810 (2007)

17. E.O. Freed, HIV-1 gag proteins: diverse functions in the virus life cycle. Virology **251**(1), 1–15 (1998)

18. D.C. Wack, W.W. Webb, Synchrotron X-ray study of the modulated lamellar phase $P_{\beta'}$ in the lecithin-water system. Phys. Rev. A **40**, 2712–2730 (1989)

19. S. Doniach, A thermodynamic model for the monoclinic (ripple) phase of hydrated phospholipid bilayers. J. Chem. Phys. **70**(10), 4587–4596 (1979)

20. M. Marder, H.L. Frisch, J.S. Langer, H.M. McConnell, Theory of the intermediate rippled phase of phospholipid bilayers. Proc. Natl. Acad. Sci. **81**(20), 6559–6561 (1984)

21. M.H. Hawton, W.J. Keeler, van der Waals energy of lecithins in the ripple phase. Phys. Rev. A **33**(5), 3333 (1986)

22. J.M. Carlson, J.P. Sethna, Theory of the ripple phase in hydrated phospholipid bilayers. Phys. Rev. A **36**(7), 3359 (1987)

23. R.E. Goldstein, S. Leibler, Model for lamellar phases of interacting lipid membranes. Phys. Rev. Lett. **61**(19), 2213 (1988)

24. W.S. McCullough, H.L. Scott, Statistical-mechanical theory of the ripple phase of lipid bilayers. Phys. Rev. Lett. **65**, 931–934 (1990)

25. K. Honda, H. Kimura, Theory on formation of the ripple phase in bilayer membranes. J. Phys. Soc. Jpn. **60**(4), 1212–1215 (1991)

26. T.C. Lubensky, F.C. MacKintosh, Theory of "ripple" phases of lipid bilayers. Phys. Rev. Lett. **71**(10), 1565 (1993)

27. K. Sengupta, V. Raghunathan, Y. Hatwalne, Role of tilt order in the asymmetric ripple phase of phospholipid bilayers. Phys. Rev. Lett. **87**(5), 055705_1–055705_4 (2001)

28. M.A. Kamal, A. Pal, V.A. Raghunathan, M. Rao, Theory of the asymmetric ripple phase in achiral lipid membranes. Europhys. Lett. **95**(4), 48004 (2011)

29. A.H. de Vries, S. Yefimov, A.E. Mark, S.J. Marrink, Molecular structure of the lecithin ripple phase. Proc. Natl. Acad. Sci. **102**(15), 5392–5396 (2005)

30. O. Lenz, F. Schmid, Structure of symmetric and asymmetric "ripple" phases in lipid bilayers. Phys. Rev. Lett. **98**, 058104 (2007)

31. H.L. Scott, Monte Carlo studies of a general model for lipid bilayer condensed phases. J. Chem. Phys. **80**(5), 2197–2202 (1984)

32. A. Debnath, K.G. Ayappa, V. Kumaran, P.K. Maiti, The influence of bilayer composition on the gel to liquid crystalline transition. J. Phys. Chem. B **113**(31), 10660–10668 (2009). PMID: 19594148

33. M.J. Janiak, D.M. Small, G.G. Shipley, Nature of the thermal pretransition of synthetic phospholipids: dimyristoyl- and dipalmitoyllecithin. Biochemistry **15**(21), 4575–4580 (1976)

34. B.R. Copeland, H.M. McConnell, The rippled structure in bilayer membranes of phosphatidyl-
    choline and binary mixtures of phosphatidylcholine and cholesterol. Biochim. Biophys. Acta
    (BBA) – Biomembr. **599**(1), 95–109 (1980)
35. D. Ruppel, E. Sackmann, On defects in different phases of two-dimensional lipid bilayers.
    J. Phys. **44**(9) 1025–1034 (1983)
36. J. Zasadzinski, M. Schneider, Ripple wavelength, amplitude, and configuration in lyotropic
    liquid crystals as a function of effective headgroup size. J. Phys. **48**(11), 2001–2011 (1987)
37. W.J. Sun, S. Tristram-Nagle, R.M. Suter, J.F. Nagle, Structure of the ripple phase in lecithin
    bilayers. Proc. Natl. Acad. Sci. **93**(14), 7008–7012 (1996)
38. J. Katsaras, S. Tristram-Nagle, Y. Liu, R. Headrick, E. Fontes, P. Mason, J.F. Nagle,
    Clarification of the ripple phase of lecithin bilayers using fully hydrated, aligned samples.
    Phys. Rev. E **61**(5), 5668 (2000)
39. K. Sengupta, V.A. Raghunathan, J. Katsaras, Structure of the ripple phase of phospholipid
    multibilayers. Phys. Rev. E **68**, 031710 (2003)
40. P.C. Mason, B.D. Gaulin, R.M. Epand, G.D. Wignall, J.S. Lin, Small angle neutron scattering
    and calorimetric studies of large unilamellar vesicles of the phospholipid dipalmitoylphos-
    phatidylcholine. Phys. Rev. E **59**(3), 3361 (1999)
41. R.A. Parente, B.R. Lentz, Phase behavior of large unilamellar vesicles composed of synthetic
    phospholipids. Biochemistry **23**(11), 2353–2362 (1984)
42. T.J. McIntosh, Differences in hydrocarbon chain tilt between hydrated phos-
    phatidylethanolamine and phosphatidylcholine bilayers. a molecular packing model.
    Biophys. J. **29**(2), 237–245 (1980)
43. J. Nagle, Theory of lipid monolayer and bilayer phase transitions: effect of headgroup
    interactions. J. Membr. Biol. **27**(1), 233–250 (1976)
44. G. Cevc, Polymorphism of the bilayer membranes in the ordered phase and the molecular
    origin of the lipid pretransition and rippled lamellae. Biochim. Biophys. Acta (BBA)-
    Biomembr. **1062**(1), 59–69 (1991)
45. R. Wittebort, C. Schmidt, R. Griffin, Solid-state carbon-13 nuclear magnetic resonance of the
    lecithin gel to liquid-crystalline phase transition. Biochemistry **20**(14), 4223–4228 (1981)
46. M.B. Schneider, W.K. Chan, W.W. Webb, Fast diffusion along defects and corrugations in
    phospholipid $P_{\beta'}$, liquid crystals. Biophys. J. **43**(2), 157–165 (1983)
47. D. Marsh, Molecular motion in phospholipid bilayers in the gel phase: long axis rotation.
    Biochemistry **19**(8), 1632–1637 (1980)
48. B.A. Cunningham, A.-D. Brown, D.H. Wolfe, W.P. Williams, A. Brain, Ripple phase formation
    in phosphatidylcholine: effect of acyl chain relative length, position, and unsaturation. Phys.
    Rev. E **58**(3), 3662 (1998)
49. M. Rappolt, G. Pabst, G. Rapp, M. Kriechbaum, H. Amenitsch, C. Krenn, S. Bernstorff,
    P. Laggner, New evidence for gel-liquid crystalline phase coexistence in the ripple phase of
    phosphatidylcholines. Eur. Biophys. J. **29**(2), 125–133 (2000)

Chapter 2
Structural and Material Perturbations of Lipid Bilayers Due to HIV-1 Tat Peptide

Abstract This chapter reports the effect on lipid bilayers of Tat, the transactivator of transcription, which is an important protein for HIV-1 infection. Synergistic use of low-angle X-ray scattering (LAXS) and atomistic molecular dynamic simulations (MD) revealed Tat peptides binding to lipid headgroups. This binding induced the local lipid phosphate groups to move closer to the center of the bilayer. The position of the positively charged guanidinium components of the arginines was also indicated. A single lipid component sample and samples consisting of mixtures of different lipids were studied. Generally, the Tat peptide decreased the bilayer bending modulus and increased the area/lipid. Although a mechanism for translation remains obscure, this study suggests that the peptide/lipid interaction makes the Tat peptide poised to translocate from the headgroup region.

2.1 Introduction

Cell penetrating peptides (CPP) easily penetrate cell membranes [1–3]. The two most extensively studied CPPs are Tat and penetratin. This chapter focuses on the transactivator of transcription, Tat, from the HIV-1 virus, which plays a role in AIDS progression. Earlier work showed that the HIV-1 Tat protein (86 amino acids) was efficiently taken up by cells, and concentrations as low as 1 nM were sufficient to transactivate a reporter gene expressed from the HIV-1 promoter [4, 5]. It has been reported that Tat protein uptake does not require ATP [6]. Studies using inhibitors of different types of endocytosis, including clathrin and caveolae-mediated, or receptor-independent macropinocytosis reached the same conclusion that ATP mediated endocytosis is not involved in Tat protein penetration [7–10]. However, this issue is controversial, as other studies found evidence for endocytosis in Tat protein import [11–19]. Still other studies have concluded that an ATP requirement for Tat protein entry depends on the size of the cargo attached to Tat protein, or on the specific cell type [20–22]. The part of the Tat protein responsible for cellular uptake was attributed to a short region, $G_{48}RKKRRQRRRPPQ_{60}$, which is particularly rich in basic amino acids [6]. Deletion of three out of eight positive charges in this region caused loss of its ability to translocate [6]. To avoid confusion, this short basic region will be called Tat_{48-60}, and the peptide used in this chapter

© Springer International Publishing Switzerland 2015
K. Akabori, *Structure Determination of HIV-1 Tat/Fluid Phase Membranes and DMPC Ripple Phase Using X-Ray Scattering*, Springer Theses,
DOI 10.1007/978-3-319-22210-3_2

(Y_{47}GRKKRRQRRR$_{57}$) will be simply called Tat. The entire amino acid sequence will be called Tat protein. Tat$_{48-60}$ was shown to be responsible for Tat protein permeation into the cell nucleus and the nucleoli [6], and this was confirmed using live cell fluorescence in SVGA cells [23]. Tat$_{48-60}$ was shown to have little toxicity on HeLa cells at $100\,\mu$M concentration [6], but Tat protein was toxic to rat brain glioma cells at $1-10\,\mu$M [24]. Interestingly, no hemolytic activity was found when human erythrocytes were incubated with a highly neurotoxic concentration ($40\,\mu$M) of the Tat protein [24]. These results prompt the question, what is the mechanism of Tats membrane translocation? To address this question, many biophysical studies have used simple model biomembranes composed of a small number of lipid types. Without proteins, there is no possibility for ATP-dependent Tat translocation, thus ruling out endocytosis if translocation occurs. For example, Mishra et al. reported that the rate of entry into giant unilamellar vesicles (GUVs) composed of phosphatidylserine (PS):phosphatidylcholine (PC) (1:4 mole ratio) lipids of rhodamine-tagged Tat is immeasurably slow, but it crosses GUVs composed of PS:PC:phosphatidylethanolamine (PE) (1:2:1) lipids within 30 s [25]. This study suggests that negative curvature, induced by the PE, facilitates translocation. In a subsequent study using large unilamellar vesicles (LUVs), which have a much smaller diameter than GUVs, Tat did not release an encapsulated fluorescent probe from LUVs composed of lipids modeling the outer plasma membrane, PC:PE:sphingomyelin:cholesterol (1:1:1:1.5) but did release the probe in LUVs composed of BMP:PC:PE (77:19:4) [26]; BMP (bis(monoacylglycero)-phosphate) is an anionic lipid specific to late endosomes. In that study [26], the inclusion of PE did not cause leaky fusion in the absence of a negatively charged lipid. The contrasting results in these two experiments may also be due to the use of LUVs instead of GUVs since it was reported that Tat does not translocate across LUVS of PC:phosphatidylglycerol(PG) (3:2) but does translocate across GUVs of the same lipid composition [27]. In a similar experiment, Tat did not translocate into egg PC LUVs [28]. In another experiment confirming these results, Tat did not translocate into GUVs containing only PC with 20 mol% cholesterol, but when PS or PE was included with PC, rapid Tat translocation was observed [29]. These experiments demonstrate that Tat translocation is influenced by both model system geometries and composition.

Some researchers have suggested that pores may form during Tat translocation. Although direct conductance measurements of Tat and lipid membranes have not been carried out, two studies measured conductance with the somewhat similar CPP, the oligoarginine R_9C peptide. Using single-channel conductance of gramicidin A in planar lipid membranes consisting of anionic, neutral, or positively charged lipids, R_9C did not increase conductance, even in anionic lipid membranes [30]. By contrast, in a similar experiment using planar lipid membranes, R_9C increased conductance in PC:PG (3:1) membranes with increasing destabilization over time [31]. Thus questions remain about Tat mediated pore formation. In the GUV experiment with Tat mentioned above [29], Ciobanasu et al., using size exclusion methods, suggested a pore in the nanometer range, which could only be passed by small dye tracer molecules. Thus, if a pore forms, it is likely to be small and transitory.

The secondary structure of Tat has been characterized by many researchers. Thoren et al. carried out circular dichroism (CD) spectroscopy on a variation of Tat where the penultimate proline on Tat_{48-60} was replaced by a tryptophan [27]. Their study found a random coil secondary structure in aqueous solution as well as when Tat_{48-60} was mixed with PC:PG:PE (65:35:5) LUVs. Ziegler et al. [10] obtained the same result using CD in PC:PG (3:1) vesicles. In addition, solid state NMR (nuclear magnetic resonance) has identified a random coil structure of Tat in DMPC:DMPG (8:7) multibilayers [32]. In the larger Tat_{1-72} protein, NMR measurements at pH 4 have determined that there is no secondary structure, with a dynamical basic region [33]. Similarly, NMR was used to study the full Tat protein and found a highly flexible basic region [34]. These previous studies indicate that an alpha helix is not required for Tat translocation ability.

Regarding the mechanism of translocation of this randomly structured, short basic peptide, many models have been proposed based on the conflicting results listed above. Molecular dynamics simulations offer some insight into the molecular details of translocation. Herce and Garcia simulated the translocation of Tat (Y_{47}GRKKRRQRRR$_{57}$) across DOPC at various lipid:peptide molar ratios [35]. Their simulations indicated that Tat binds to the phosphate headgroups, with 1 Tat binding with 14 lipids, each positive charge on Tat associated with nearly 2 phosphate groups [35]. Translocation involved a localized thinning, and snorkeling of arginine side chains through the hydrophobic layer to interact with phosphates on the other side of the membrane. This allowed some water molecules to penetrate the membrane along with Tat, forming a pore [35]. In this simulation, performed without inclusion of counterions, pore formation was only observed at high ratios of peptide:lipid (1:18) or at elevated temperature. However, a subsequent Gromacs simulation with counterions found no thinning and no pore formation when Tat was added to DOPC membranes [36]. Instead they found a membrane invagination associated with a cluster of Tat peptides. From their findings, the authors suggested that Tat translocation occurs via micropinocytosis [36].

In this thesis, I combined experimental low-angle X-ray scattering (LAXS) data with MD simulations from our collaborators to obtain the structure of fully hydrated, oriented lipid bilayers with Tat added at several mole ratios. The lipid systems were DOPC, DOPC:DOPE (3:1 mole ratio), DOPC:DOPE (1:1), DOPC:DOPS (3:1), and a mimic of the nuclear membrane (POPC:POPE:POPS:SoyPI:Chol, 69:15:2:4:11 (mole ratio)).

2.2 Materials and Methods

2.2.1 Stock Solutions

Synthesized lipids were purchased from Avanti Polar Lipids (Alabaster, AL). Membrane mimics for Tat experiments were prepared by first dissolving lyophilized lipids in chloroform and then mixing these stock solutions to create the lipid compositions DOPC, DOPC:DOPE (3:1), DOPC:DOPE (1:1), DOPC:DOPS (3:1) and nuclear membrane mimic (POPC:POPE:POPS:SoyPI:Cholesterol, 69:15:2:4:11) (based on Ref. [37]). Peptide (Y_{47}GRKKRRQRRR$_{57}$) was purchased in three separate lots from the Peptide Synthesis Facility (University of Pittsburgh, Pittsburgh, PA); mass spectroscopy revealed greater than 95 % purity. This Tat peptide corresponds to residues (47–57) of the 86 residues in the Tat protein [6]. Tat was dissolved in HPLC trifluoroethanol (TFE) and then mixed with lipid stock solutions in chloroform to form mole fractions between 0.0044 and 0.108. The weight of Tat in these mole fractions was corrected for protein content (the remainder being 8 trifluoroacetate counterions from the peptide synthesis). Solvents were removed by evaporation in the fume hood followed by 2 h in a vacuum chamber at room temperature.

2.2.2 Thin Film Samples

For Tat experiments, 4 mg of a dried lipid/peptide mixture in a glass test tube was re-dissolved in HPLC chloroform:TFE (2:1 v:v) for most of the lipid compositions. DOPC:DOPS (3:1) mixtures required chloroform:hexafluoroisopropanol(HIP) (1:1 v:v) in order to solubilize the negatively charged DOPS. 200 µl of 4 mg mixtures in solvents were plated onto silicon wafers (15 × 30 × 1 mm) via the rock and roll method [38] to produce stacks of ∼1800 well-aligned bilayers; solvents were removed by evaporation in the fume hood, followed by 2 h under vacuum. Samples were prehydrated through the vapor in polypropylene hydration chambers at 37 °C for 2 to 6 h directly before hydrating in the X-ray hydration chamber [39] for 0.5 to 1 h.

2.2.3 Volume Measurements

Multilamellar vesicles (MLVs) were prepared by mixing dried lipid and Tat mixtures with MilliQ water to a final concentration of 2–5 wt% in nalgene vials and cycling three times between 20 and 60 °C for 10 min at each temperature with vortexing. Pure Tat was dissolved in water at 0.4 wt%.

Table 2.1 Amino acid data. To calculate the molecular weight of Tat, subtract 18 for each water that gets removed by hydrolysis when forming a peptide backbone

Code	Amino acid	Chemical formula	Molecular weight (g/mol)
K	Lysine	$C_6H_{14}N_2O_2$	146.2
R	Arginine	$C_6H_{14}N_4O_2$	174.2
G	Glycine	$C_2H_5NO_2$	75.1
Y	Tyrosine	$C_9H_{11}NO_3$	181.2
Q	Glutamine	$C_5H_{10}N_2O_3$	146.1

Volumes of lipid mixtures with and without peptides in fully hydrated MLVs were determined at 37 ± 0.01 °C using an Anton-Paar USA DMA5000M (Ashland, VA) vibrating tube densimeter. This instrument measures the average density of a solution ρ_s and compares it to the density of air ρ_0 using $\rho_s - \rho_0 = k(\tau_s - \tau_0)^2$ where k is an instrumental constant that depends on the atmospheric pressure.

The Tat peptide sequence used in X-ray experiments and MD simulations was Y_{47}GRKKRRQRRR$_{57}$. Table 2.1 lists the chemical formulas and molecular weights of the pertinent amino acids for convenience. The molecular weight of this sequence is 1560 g/mol. The Tat peptides were synthesized in trifluoroacetic acid, CF_3CO_2H, and were made into a powder form by the freeze-dry method. Therefore, each positively charged amino acid, such as an arginine and lysine, was counter-balanced by a trifluoroacetate (TFA, $C_2F_3O_2$). Since Tat has six arginines and two lysines, it was counter-balanced by eight TFAs. The peptide-counterions complex has a molecular weight of $1560 + 113 \times 8 = 2464$. We used the molecular weight of this complex in order to calculate the molarity of Tat correctly. The same molecular weight was also used in preparing oriented samples.

The Tat volume V_{Tat} was calculated from the measured average density of a Tat-water solution. The partial specific volume of water in a system with excess water is the same as the volume of bulk water. Then the density of a Tat-water solution is equal to the mass of a Tat-water solution divided by the sum of the volumes of water and Tat,

$$\rho_{sol} = \frac{m_w + m_c}{V_w + V_c N_c},$$ (2.1)

where m_w and m_c are the total masses of water and a Tat-TFA complex, respectively, V_w is the total volume of water, V_c is the molecular volume of a Tat-TFA complex, and N_c is the total number of complexes in the solution. Defining $V_w = m_w/\rho_w$ and $N_c = N_A m_c/W_c$, where W_c is the molecular weight of the complex, N_A is Avogadro's number, and ρ_w is the density of water, we have

$$V_c = \frac{W_c}{\rho_{sol} N_A} \left(1 + \frac{m_w}{m_c} \left(1 - \frac{\rho_{sol}}{\rho_w} \right) \right),$$ (2.2)

which allows us to calculate the molecular volume of a Tat-TFA complex from the experimentally measured quantities. Assuming that the molecular volume scales with the molecular weight, we have $V_{\text{Tat}} = 1560/2464 \times V_c\ \text{Å}^3$.

2.2.4 X-Ray Setup

Figure 2.1 shows a schematic of our X-ray setup omitting details of the flightpath upstream of the sample hydration chamber. MilliQ water filled the bottom of the hydration chamber, providing water vapor for the sample. The sample holder was mounted on a rotation motor, which allowed continuous rotation of the sample during an X-ray exposure for low angle X-ray scattering (LAXS) as well as fixed angles of incidence ω for wide angle X-ray scattering (WAXS). A Peltier cooling/heating element was attached to the sample holder, and the sample was situated on top of the Peltier element. Using the Peltier the sample hydration level was adjusted by maintaining a temperature difference between the sample and water vapor. The hydration chamber walls were made of aluminum within which water at a constant temperature T circulated to provide a thermal bath: $T = 37\,°\text{C}$ for Tat experiments and $18\,°\text{C}$ for ripple phase experiments. Entrance and exit

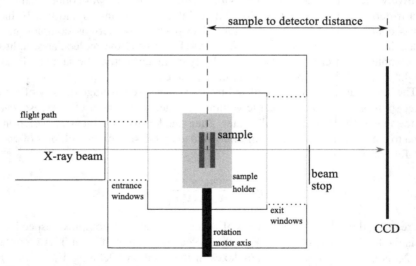

Fig. 2.1 Schematic diagram of a top view of the X-ray setup for LAXS experiments. A *lipid* sample deposited on a *dark gray* Si wafer was situated on top of a Peltier cooling/heating device, which was attached to the *light gray* sample holder. The sample holder was mounted on the *black* rotation motor axis, which provided precise control of the incident angle ω. Thin mylar windows allowed incoming and outgoing X-rays shown as a *arrow* to go through the chamber. Thin pieces of molybdenum attenuated the direct beam to avoid saturation of CCD pixels and conveniently allowed beam profile measurements due to its transparency. The flightpath and hydration chamber were filled with helium to reduce air scattering

windows for the X-ray beam were made of mylar, which caused strong mylar scattering in the wide angle region as described in Chap. 3. Additional hydration chamber details are described in [39]. The sample to detector distance was measured by indexing the standard silver behenate diffraction pattern whose D-spacing is 58.367 Å. The hydration level of a sample was estimated by measuring the average interbilayer distance, D-spacing, which was easily calculated by indexing the out-of-plane diffraction peaks using the tview software developed by Dr. Yufeng Liu. Molybdenum between the sample and the charge-coupled device (CCD) detector attenuated the direct beam; otherwise the direct beam would saturate the CCD pixels. Data reduction and correction for a CCD detector are described in detail in [40].

2.2.5 Analysis of Diffuse Scattering

Figure 2.2 shows our typical LAXS data from oriented stacks of fluctuating bilayers in the fluid phase. The analysis of diffuse scattering intensity patterns like the one shown in Fig. 2.2 yields material parameters such as the bending modulus K_c and bulk modulus B as well as the absolute form factor $|F(q_z)|$. The X-ray form factor $F(\mathbf{q})$ is the Fourier transform of the bilayer electron density profile $\rho(z)$ normal to the membrane plane and is related to the internal structure of the bilayers including Tat peptides.

The form factor $|F(q_z)|$ is obtained through the relation $I(\mathbf{q}) = S(\mathbf{q})|F(q_z)|^2$, where $I(\mathbf{q})$ is the measured intensity, and $S(\mathbf{q})$ is the structure factor. Here, the X-ray momentum transfer $\mathbf{q} = (q_r, q_z)$, indicating that the system is isotropic in-plane.

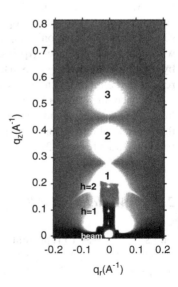

Fig. 2.2 LAXS of DOPC:DOPE (1:1) with Tat mole fraction $x_{Tat} = 0.034$ at 37 °C. White lobes of diffuse scattering intensity have large grey numbers, while lamellar orders and beam are shown to the left of the molybdenum beam attenuator (*short, dark rectangle*). q_z and q_r are the cylindrical coordinates of the sample q-space, where the q_z-axis is along the bilayer normal and the q_r-axis is along the in-plane direction. The lamellar repeat spacing was $D = 66.2$ Å

Fig. 2.3 Schematic of an oriented stack of lipid bilayers. *Thick curves* represent an instance of thermally fluctuating bilayers. The *dashed lines* show the thermally averaged positions $z = nD$ of the centers of each bilayer and $u_n(x, y)$ gives the instantaneous deviation from the average. Each bilayer extends in the $\mathbf{r} = (x, y)$ plane

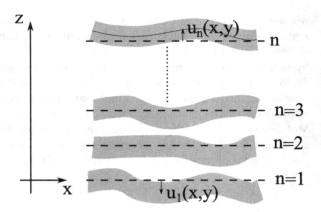

In fully hydrated multilamellar samples, $S(\mathbf{q})$ is not a sum of delta functions because of thermal fluctuations of bilayers. Calculating $S(\mathbf{q})$ requires a model free energy for bilayer fluctuations, from which the scattering pair correlation function is derived. A basic scattering theory, then, relates the scattering intensity $I(\mathbf{q})$ to the pair correlation function. For modeling the membrane fluctuations of a multilamellar system, the smectic liquid crystal free energy functional in the discrete form,

$$F = \frac{1}{2} \int d\mathbf{r} \sum_{n=0}^{N-1} \left\{ K_c \left[\nabla_r^2 u_n(\mathbf{r}) \right]^2 + B \left[u_{n+1}(\mathbf{r}) - u_n(\mathbf{r}) \right]^2 \right\}, \qquad (2.3)$$

has been shown to be adequate [41]. Here, $u_n(\mathbf{r})$ is the spatial deviation of the center of the n-th bilayer from its average position in the z direction at the in-plane location $\mathbf{r} = (x, y)$ (Fig. 2.3). The first term is the bending free energy proportional to the curvature squared with the proportionality given by a bending modulus K_c, and the second term is a harmonic approximation to the interactions between membranes with a modulus B. Once $S(\mathbf{q})$ is calculated from Eq. (2.3), $|F(q_z)|$ can be calculated by dividing the intensity by $S(\mathbf{q})$. Getting the best fit of a model $S(\mathbf{q})$ to the intensity results in the material parameters K_c and B.

We used software called NFIT developed by Dr. Yufeng Liu [41–43] to analyze the diffuse scattering and obtain K_c, B, and $|F(q_z)|$. Details of the analysis, including the theoretical derivation of $S(\mathbf{q})$ from Eq. (2.3) and its numerical computation, are found in Liu's thesis [43].

2.2.6   Modeling the Bilayer Structure

The simplest way to represent the results of X-ray data in real space is to Fourier transform the $F(q_z)$ form factors to obtain electron density profiles $\rho(z)$. However, the $\rho(z)$ so obtained are on an arbitrary scale. Furthermore, no information is obtained regarding the location of component groups of the lipid or the location of added peptides. Finally, Fourier reconstruction requires knowing the phase factors of individual reflections; this latter concern is alleviated when diffuse scattering is obtained as the zeros in $I(q_z)$ locate where the phase factors change sign. Modeling uses the intensities, not the phase factors, obtains absolute electron densities, and estimates where the different components of the system are located. Early so-called strip models used constant $\rho(z)$ in different z regions [44]. This has been improved by using error functions to smear the artificially sharp edges of the strip model [45, 46]. When the width of two error function interfaces are wide compared to the distances between the edges, the profile becomes a Gaussian. Models consisting of sums of Gaussians have been used [47]. A hybrid model used positive Gaussians for the headgroup and a negative Gaussian for the terminal methyl region superimposed on a modulated baseline for the water and the hydrocarbon [48], which was later replaced by error functions for the hydrocarbons and for water [49]. This lab now uses the SDP method which imposes a volumetric constraint to account for the water profile [50].

The SDP method is applicable to joint fitting of neutron and X-ray scattering data when a particular parsing of the component groups is employed. For X-ray scattering data alone, a different parsing is more appropriate. The parsing of DOPC into molecular components is shown in Fig. 2.4. The phosphate/choline (PC) and carbonyl/glycerol (CG) components together make up the lipid headgroup whereas the hydrocarbon chain region (HC) is divided into two components, the methylene (CH_2) and methine (CH) group combination (denoted as CH_2+CH) and terminal methyl groups (CH_3). We combine methylene (CH_2) and methine groups (CH) in order to minimize the number of fitting parameters.

2.2.6.1 Functional Forms

Our model for the electron density profile (EDP) of the Tat/lipid bilayer system consists of five structural subgroups: PC, CG, CH_2+CH, CH_3, and Tat (see Fig. 2.5). Assuming bilayers are centrosymmetric, the volume probability distributions P_i of components PC, CG, CH_3, and Tat are described by Gaussian functions,

$$P_i(z) = \frac{c_i}{\sqrt{2\pi}} \left(\exp\left\{ -\frac{(z+z_i)^2}{2\sigma_i^2} \right\} + \exp\left\{ -\frac{(z-z_i)^2}{2\sigma_i^2} \right\} \right), \qquad (2.4)$$

where i specifies a particular molecular component, PC, CG, Tat, $CH+CH_2$, and CH_3, c_i is an integrated area underneath the curve, σ_i is the width, z_i is the center, and the two parts of the expression describe the two bilayer leaflets.

The hydrocarbon chain region (HC) is represented by error functions,

$$P_{HC}(z) = \frac{1}{2}[\text{erf}(z, -z_{HC}, \sigma_{HC}) - \text{erf}(z, z_{HC}, \sigma_{HC})], \qquad (2.5)$$

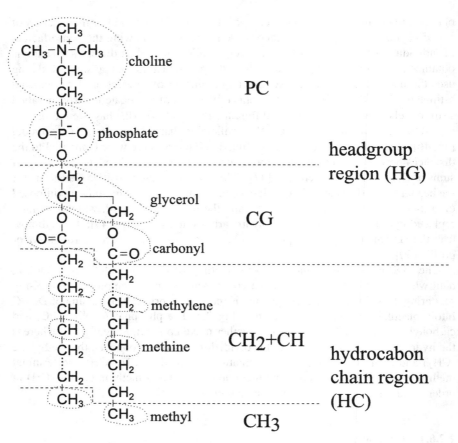

Fig. 2.4 Schematic of DOPC showing each molecular component. The *dashed lines* show where the lipid is divided into different components. The lipid headgroup is divided into two components, phosphate-choline (PC) and carbonyl-glycerol (CG). The hydrocarbon chain region is also divided into two components, methylene+methine (CH$_2$+CH) and terminal methyl groups (CH$_3$). Each hydrocarbon chain has 18 carbons. Repeated methylene groups are shown by *dots*

where

$$\mathrm{erf}(z, z_i, \sigma_i) = \frac{2}{\sqrt{\pi}} \int_0^{\frac{z-z_i}{\sqrt{2}\sigma}} dx \, e^{-x^2}. \tag{2.6}$$

The volume probability distribution for the methylene and methine group combination can then be expressed as

$$P_{\mathrm{CH}_2+\mathrm{CH}}(z) = P_{\mathrm{HC}}(z) - P_{\mathrm{CH}_3}(z). \tag{2.7}$$

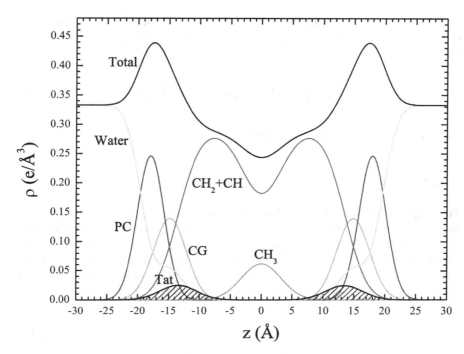

Fig. 2.5 A model electron density profile for DOPC with Tat. Lipid components are defined in Fig. 2.4. Tat profile is the *black shaded curve*. The *black solid line* labelled 'Total' is the sum of all components

This definition enforces the total probability P_{HC} in the hydrocarbon chain region to equal one, which in turn means that placement of Tat in the chain region is prohibited. We call the model defined by Eq. (2.7) Tat-in-headgroup (THG). To allow Tat to be placed inside the hydrocarbon chain region, we also consider an alternative definition,

$$P_{CH_2+CH}(z) = P_{HC}(z) - P_{CH_3}(z) - P_{Tat}(z), \tag{2.8}$$

where the volume probability of the CH_2+CH combined component is reduced by the Tat volume probability distribution. We call this model Tat-in-hydrocarbon-chain (THC). The spatial conservation requires the water volume probability distribution to be

$$P_W(z) = 1 - P_{PC}(z) - P_{CG}(z) - P_{Tat}(z) - P_{HC}(z) \tag{2.9}$$

for THG and

$$P_W(z) = 1 - P_{PC}(z) - P_{CG}(z) - P_{HC}(z) \tag{2.10}$$

for THC.

Because X-rays measure the contrast between the bilayer and surrounding solvent, the experimental form factor is compared to the water subtracted model form factor,

$$F(q_z) = 2 \int_0^{\frac{D}{2}} dz \left(\sum_i (\rho_i - \rho_W) P_i(z) \right) \cos(q_z z), \qquad (2.11)$$

where i = PC, CG, Tat, CH+CH$_2$, and CH$_3$.

2.2.6.2 Constraints

The height of the hydrocarbon chain error function is fixed to one by imposing spatial conservation, whereas the mean position of the terminal methyls is fixed to $z_{CH_3} = 0$ by symmetry arguments. The total lipid volume V_L is fixed to the experimentally measured value. The headgroup volume V_{HL} was determined to be 331 Å^3 for gel phase phosphatidylcholine bilayers [51], and we assume the same volume for the fluid phase phosphatidylcholine bilayers. The volumes of PC and CG components satisfy

$$V_{PC} + V_{CG} = V_{HL}, \qquad (2.12)$$

and the volumes of CH$_3$ and CH$_2$+CH components satisfy

$$2 (16 V_{CH_2+CH} + V_{CH_3}) = V_L - V_{HL}. \qquad (2.13)$$

These component volumes constrain the heights of the Gaussians as

$$c_{PC} = \frac{V_{PC}}{A_L \sigma_{PC}} \qquad (2.14)$$

$$c_{CG} = \frac{V_{CG}}{A_L \sigma_{CG}} \qquad (2.15)$$

$$c_{CH_3} = \frac{2 V_{CH_3}}{A_L \sigma_{CH_3}} \qquad (2.16)$$

$$c_{Tat} = \frac{V_{Tat}}{A_L \sigma_{Tat}} \qquad (2.17)$$

where A_L is area per lipid.

The ratio of the carbonyl/glycerol volume to the headgroup volume V_{HL} was reported to be 0.41 [52], so we constrain the CG component volume to 135.7 Å^3 and the PC component volume to 195.3 Å^3.

The most detailed structural study on DOPC to date was published by Braun et al. [52], and many of the constraints on our model parameters can be derived from their study. However, in that work, the authors used the SDP model [50], which is specifically tailored for simultaneous analysis of neutron and X-ray form factors. Therefore, we need to convert their structural results to the corresponding parameters in our simpler X-ray model. For example, from the reported values of the ratio of the volumes of the chain terminal methyl (CH_3) to the chain methylenes (CH_2) and the ratio of the volumes of the chain methines (CH) to the chain methylenes, we can calculate the ratio r_{CH_3} of the volumes of CH_3 to the CH_2 and CH combined component. Furthermore, the study by Braun et al. was at 30 °C while our study was at 37 °C, so our measured volume of DOPC was slightly greater.

At 30 °C, the volume of DOPC was reported to be 1303 Å^3 [50], so the volume of hydrocarbon chain region at the same temperature is $1303 - 331 = 972$ Å^3. The ratio r of the volumes of the chain terminal methyl (CH_3) to the chain methylenes (CH_2) was reported to be 1.95, and the ratio r_{12} of the volumes of the chain methines (CH) to the chain methylenes was 0.91 at 30 °C. Because there are 14 CH_2 groups, 2 CH groups, and 1 CH_3 group in each DOPC hydrocarbon chain, we have $2 \times (14V_{CH_2} + 2V_{CH} + V_{CH_3}) = 972$ Å^3. Using $r = V_{CH_3}/V_{CH_2} = 1.95$ and $r_{12} = V_{CH}/V_{CH_2} = 0.91$, we get $V_{CH_2} = 27.3$ Å^3, $V_{CH} = 24.9$ Å^3, and $V_{CH_3} = 53.3$ Å^3. These calculated volumes lead to $V_{CH_3}/V_{CH_2+CH} = 1.97$ for 30 °C.

At 37 °C, the volume of DOPC was measured to be 1313.5 Å^3, so we have $2 \times (16V_{CH_2+CH} + V_{CH_3}) = 1313.5 - 331$. Assuming that the ratio V_{CH_3}/V_{CH_2+CH} at 37 °C is the same as that at 30 °C gives $V_{CH_2+CH} = 27.3$ Å^3 and $V_{CH_3} = 53.9$ Å^3. We constrain the components for the hydrocarbon chain region in our model to these calculated values (Tables 2.2, 2.3, 2.4).

2.2.6.3 Fits with Lower Bounds

Modeling of the bilayer structure was done using the SDP software, as described in Sect. 2.2.6. To allow model parameters with upper and lower bounds, the SDP software was modified following the MINUIT User's Guide, section 1.3 [53].

Table 2.2 Number of electrons and volume per lipid

Lipid	Number of electrons	Volume (Å^3)
DOPC	434	1313.5
DOPE	410	1212.3
DOPC:DOPE (3:1)	428	1288.2

Table 2.3 Structural parameters for each component. n_i^e is the number of electrons and ρ_i is the average electron density. V_i is the molecular volume

Component	n_i^e	V_i (Å^3)	ρ_i (e/Å^3)
PC	97	195.3	0.497
PE	73	94.1	0.776
PC:PE (3:1)	91	170	0.535
CG	67	135.7	0.494
CH_2+CH	7.875	27.3	0.288
CH_3	9	53.9	0.167

Table 2.4 Tat basic structural parameters. The notations are the same as in Table 2.3. x_{Tat} is Tat mole fraction

Number of electrons	838	Mole fraction (x_{Tat})	n^e_{Tat}	V_{Tat} (Å^3)
Volume (Å^3)	1877	0.016	13.6	30.5
ρ_{Tat} (e/Å^3)	0.446	0.034	29.5	66.1
		0.059	53.0	118.8

Briefly, the modified minimization routine "sees" internal variables at each iteration. These internal variables can take on any values between $-\infty$ to $+\infty$, which is an assumption made in a typical minimization routine such as the simplex method and Levenburg-Marquadt algorithm. A model parameter with both lower and upper bounds (a and b, respectively) is related to its corresponding internal variable by the following transformation,

$$P_{int} = \arcsin\left(2\frac{P_{ext} - a}{b - a} - 1\right) \tag{2.18}$$

$$P_{ext} = a + \frac{b - a}{2}(\sin P_{int} + 1), \tag{2.19}$$

where P_{int} is the value of an internal variable and P_{ext} is the value of a model parameter. It is easy to show that P_{ext} can only take values between a and b. The goodness of a fit χ^2 is then calculated by transforming the internal variables to their respective model parameters via Eq. (2.19). For variables with a lower bound a only, the transformation is

$$P_{int} = \sqrt{(P_{ext} - a + 1)^2 - 1} \tag{2.20}$$

$$P_{ext} = a - 1 + \sqrt{P_{int}^2 + 1}, \tag{2.21}$$

and for variables with an upper bound b only,

$$P_{int} = \sqrt{(b - P_{ext} + 1)^2 - 1} \tag{2.22}$$

$$P_{ext} = b + 1 - \sqrt{P_{int}^2 + 1}. \tag{2.23}$$

This nonlinear transformation between internal variables and model parameters allowed model parameters with upper and lower bounds in the SDP program.

2.2.7 Molecular Dynamics Simulation

This section describes the MD simulations performed by Dr. Kun Huang, who was a graduate student of Prof. Angel Garcia at Rensselaer Polytechnic Institute when he collaborated with the Nagle/Tristram-Nagle lab. My contribution to the MD simulations was to help analyze the results.

Systems with different DOPC/Tat mole ratios (128:0, 128:2 and 128:4, corresponding to 0, 0.015, and 0.030 mole fractions) were simulated atomistically using the Gromacs 4.6.1 package [54]. DOPC was modeled by the Slipid force field [55, 56], and HIV-1 Tat was modeled by Amber 99SB [57]. Tip3p water was used [58]. The number of Tats was divided equally on each side of the bilayer to mimic experimental conditions. All systems were simulated at 310 K with a constant area in the x-y plane and 1 atm constant pressure in the z direction. Each system was simulated for 100 ns, and the last 50 ns was used as the production run. At each DOPC/Tat mole ratio, we studied systems with three different area/lipid (A_L). For the DOPC system, we fixed $A_L = 68, 70, 72 \, \text{Å}^2$; DOPC/Tat (128:2), we fixed $A_L = 72, 74, 76 \, \text{Å}^2$; DOPC/Tat (128:4), we fixed $A_L = 72, 74, 76 \, \text{Å}^2$. These areal values were based on the analysis of experimentally obtained form factors, which is discussed in Sect. 2.4.4. For systems with Tat, chloride ions were used as counterions. For each DOPC/Tat system at fixed A_L, we then conducted seven independent simulations with the center of mass (COM) of each Tat constrained at different distances from the bilayer center (18, 16, 14, 12, 10, 8, and 5 Å). In total, 45 independent simulations were conducted. The goal of the constrained simulations was to find the best match between experimental and MD simulation form factors. Comparison to the X-ray form factors was performed using the SIMtoEXP software written by Dr. Norbert Kučerka [59].

All simulations were conducted with a 2 fs time integration step. SETTLE [60] was used to constrain water molecules, and LINCS [61] was used to constrain all other bond lengths in the system. Van der Waals interactions were truncated at 1.4 nm with a twin-range cutoff scheme and a dispersion correction was applied to both energy and pressure. Electrostatic interactions were treated with the particle-mesh Ewald (PME) method [62]. The direct term for electrostatics was evaluated within 1.0 nm cutoff and the Fourier term was evaluated with a 0.12 nm grid spacing and a 4th order interpolation. Each system was simulated at 310 K using the V-rescale algorithm [63] with a 0.2 ps time coupling constant. The semi-isotropic parrinello-rahman barostat [64] was used to couple the system at 1 atm in the z direction with a 5 ps time coupling constant, while the projected area at the $x - y$ plane was fixed by setting the system compressibility to 0. We inserted the Tats into the system by initially turning off all interactions between Tats and the rest of the system, with Tats constrained at different depths. Then we slowly turned on the interactions to normal strength through thermodynamics integrations. We used umbrella potentials to constrain Tats at desired depths with a force constant of 3000 kJ/mol/nm^2.

The center of mass (COM) distance between each peptide and the bilayer was constrained by an umbrella potential. Essentially, this potential acts as a spring, where its potential energy depends on the deviation of the distance between the center of mass of Tat and DOPC from a preferred value, z_0,

$$U(z_1^{\text{Tat}}, \ldots, z_1^{\text{DOPC}}, \ldots) = -\frac{1}{2}k(z_{\text{cm}}^{\text{Tat}} - z_{\text{cm}}^{\text{DOPC}} - z_0)^2.$$

Then, $-\partial U/\partial z_i$ is the external force acting on atom, i.

2.3 Analysis of Molecular Dynamics Simulation Data

2.3.1 SIMtoEXP Program

This section briefly describes the SIMtoEXP program developed by Kučerka et al. [59]. Essentially, for each snapshot, the positional distribution of each atom averaged over the xy plane is calculated. Then, the distribution is averaged over snapshots. The product of this distribution and the average electron density gives the electron density profile of the atom. The sum over all the electrons provides the total electron density profile. This total electron density profile minus the average electron density of water is Fourier transformed to provide the X-ray form factor.

$$F^{\text{sim}}(q_z) = \int_0^\infty dz(\rho(z) - \rho_{\text{W}})\cos(q_z z). \tag{2.24}$$

Simulated electron density profiles were symmetrized, and then $F^{\text{sim}}(q_z)$ was calculated with $\rho_{\text{W}} = 0.326$ e/Å^3, which was the average electron density of water molecules in the MD simulations. Because $\rho(z)$ is equal to ρ_{W} outside the bilayer, the upper integration limit can be truncated to a finite value.

Because the experimental form factor is in arbitrary units, it is scaled by a single constant a to produce the best fit to the simulated form factor through a linear least squares fit that minimizes the following goodness of fit

$$\chi^2 = \sum_i \left(\frac{1}{\sigma_i}\left(a|F_i^{\text{exp}}| - |F^{\text{sim}}(q_{z,i})|\right)\right)^2 \tag{2.25}$$

where σ_i is the input experimental uncertainty and F_i^{exp} is the experimental form factor measured at $q_z = q_{z,i}$. χ^2 defined by Eq. (2.25) does not keep the relative errors $\sigma_i/|F_i^{\text{exp}}|$ constant. To properly calculate the goodness of a fit, relative errors must be independent of an overall scaling factor a, so the χ^2 values calculated by the program were multiplied by $1/a^2$. These corrected χ^2 values are reported in this chapter.

2.3.2 Local Thinning of Membranes

The SIMtoEXP program gives the average quantities for each leaflet. Our X-ray data are only sensitive to the average bilayer electron density; in contrast, local information concerning Tat-bilayer interactions can be obtained from MD simulations. In this section, we discuss a method to extract a local membrane thickness around the Tat peptides from the MD simulation trajectories.

One of the expected effects of Tat interacting with a bilayer is compression of the lipid bilayer along the z-direction. It is reasonable to assume that this compression is greater near Tat and weaker far from Tat. Then, the distance D_{phos} between phosphorus atoms in opposite leaflets near Tat should be different from the distance between phosphorus atoms away from Tat. For a small Tat concentration, D_{phos} is the same as that of pure DOPC if the distance from all Tats is large enough. For our experimental concentrations, the thinning effect may extend throughout the bilayer because the lateral effect of Tat might have a larger lateral decay length than the distance between Tats. Whether that is the case or not, we expect that the bilayer thickness near a Tat is smaller than the average thickness, so D_{phos} should represent the actual thinning effect due to Tat.

First, let us define what we mean by lipids close to Tat. As in Fig. 2.6, we imagine a cylinder around Tat and find all the phosphorus atoms within it. Approximating Tat as a cylinder with its height H_{Tat} given by the FWHM of simulated electron density distribution, its radius R_{Tat} is calculated from the experimentally determined volume $V_{Tat} = 1876\,\text{Å}^3$ using $R_{Tat} = \sqrt{V_{Tat}/(\pi H_{Tat})}$. Let us define the lateral center of the cylinder as the center of mass of each Tat. Then we define D_{phos} using only

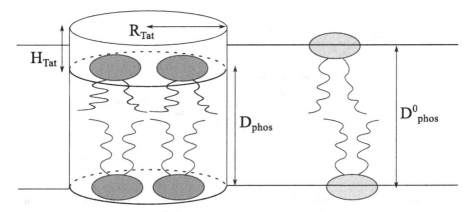

Fig. 2.6 Our simple model to extract the local bilayer thickness from simulation trajectories. Tat is modeled as a cylinder with height H_{Tat} and radius R_{Tat}. The local bilayer thickness is defined as D_{phos}. The thickness of the unperturbed DOPC bilayer is D_{phos}^0. Four lipids on the left fall within the imaginary cylinder extended from Tat. Two unperturbed lipids are on the right.

those lipids whose phosphorus atoms lie within these R_{Tat} cylinders around a Tat. Then $D_{phos} = z^+_{phos} - z^-_{phos}$ where z^+_{phos} and z^-_{phos} are the average z of the n^+ (n^-) lipids in the upper and lower monolayer, respectively.

The algorithm for doing the above was straightforward. For each time frame, the positions (x_i, y_i, z_i) of each Tat, i, are listed. We chose phosphorus atoms whose (x, y) lateral position lay within R_{Tat} of any Tat position. Then, z positions of the chosen phosphorus atoms were placed in a list, from which z_{phos} was calculated. Dr. Huang supplied us with files containing the value of z_{phos} at each snapshot, and I wrote a script to average over many snapshots to improve statistics.

2.3.3 Lateral Decay Length of Membrane Thinning

This section describes a method to measure the lateral decay length of membrane thinning due to Tat-lipid interactions. As in the previous section, Tat is modeled as a cylinder with its radius equal to R_{Tat}, height H_{Tat}, and volume V_{Tat} such that $R_{Tat} = \sqrt{V_{Tat}/(\pi H_{Tat})}$. Let $h(r)$ represent the phosphorus height profile of a leaflet as in Fig. 2.7. The two leaflets are assumed to be decoupled. In our model, lipids are separated into three regions: suppressed, boundary, and unperturbed region. The suppressed region extends from $r = 0$ to R_{Tat} and is directly beneath (above) Tat in the top (bottom) leaflet. In this region, lipids are uniformly compressed by Tat

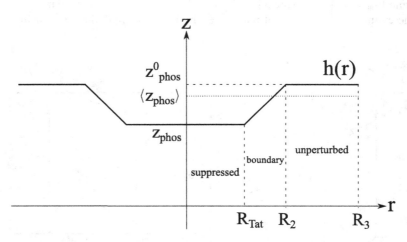

Fig. 2.7 Simple model of the lateral decay of the membrane thickness perturbation due to Tat. The suppressed region is for $0 \le r < R_{Tat}$, the boundary region for $R_{Tat} \le r < R_2$, and the unperturbed region for $R_2 \le< R_3$. z_{phos} is the average z position of phosphorus atoms measured from the bilayer center within the suppressed region. z_{phos} was obtained directly from the MD simulation trajectories as described in Sect. 2.3.2. z^0_{phos} is the average z position of phosphorus atoms measured from the bilayer center in the unperturbed region. $\langle z_{phos} \rangle = \langle h(r) \rangle$ is half of the D_{phos} distance averaged over all lipids

toward the center of the bilayer, so that $h(r)$ is a constant equal to z_{phos}. From $r = R_{Tat}$ to R_2 is the boundary region, where $h(r)$ is assumed to linearly increase with the lateral distance r. The lateral decay length of membrane thinning is given by R_2. In the unperturbed region ($r > R_2$), lipids do not interact with Tat, behaving identically to DOPC, so the phosphorus position is the same as that of DOPC. A continuous $h(r)$ that satisfies the above criteria is

$$h(r) = \begin{cases} z_{phos} & \text{if } 0 \leq r < R_{Tat} \\ mr + b & \text{if } R_{Tat} \leq r < R_2 \\ z_{phos}^0 & \text{if } R_2 \leq r < R_3 \end{cases} \tag{2.26}$$

with $m = (z_{phos} - z_{phos}^0)/(R_{Tat} - R_2)$ and $b = (z_{phos}^0 R_{Tat} - z_{phos}R_2)/(R_{Tat} - R_2)$. Approximating the simulation box as a cylinder gives $R_3 = \sqrt{NA_L/\pi}$, where N is the number of lipids in a leaflet. z_{phos} can be measured directly from simulation trajectories. z_{phos}^0 is half of the average D_{phos} in a DOPC simulation, which can be easily obtained from the SIMtoEXP program. The average height profile over the monolayer, $\langle h(r) \rangle$, also can be obtained from the program in the same manner. The only unknown is R_2.

Let us calculate $\langle h(r) \rangle$. In cylindrical coodinates,

$$\langle h(r) \rangle = \frac{1}{\pi R_3^2} \int_0^{2\pi} d\phi \int_0^{R_3} dr \, rh(r). \tag{2.27}$$

The ϕ integration is trivial. The r integration is

$$\int_0^{R_3} dr \, rh(r)$$

$$= \int_0^{R_{Tat}} dr \, z_{phos}r + \int_{R_{Tat}}^{R_2} dr(mr + b)r + \int_{R_2}^{R_3} dr \, z_{phos}^0 r$$

$$= \frac{1}{2} \left[z_{phos}R_{Tat}^2 + z_{phos}^0 (R_3^2 - R_2^2) \right] + \frac{1}{3}m \left(R_2^3 - R_{Tat}^3 \right) + \frac{1}{2}b \left(R_2^2 - R_{Tat}^2 \right)$$

$$= \frac{1}{2} \left[z_{phos}R_{Tat}^2 + z_{phos}^0 (R_3^2 - R_2^2) \right] + \frac{1}{3} \left(z_{phos}^0 - z_{phos} \right) \left(R_2^2 + R_{Tat}R_2 + R_{Tat}^2 \right)$$

$$+ \frac{1}{2} \left(z_{phos}R_2 - z_{phos}^0 R_{Tat} \right) (R_{Tat} + R_2). \tag{2.28}$$

Using Eq. (2.28), we get

$$\langle h(r) \rangle = \frac{\left(z_{phos} - z_{phos}^0 \right) \left(R_{Tat}^2 + R_{Tat}R_2 + R_2^2 \right) + 3z_{phos}^0 R_3^2}{3R_3^2}. \tag{2.29}$$

Equation 2.29 is a quadratic equation in terms of R_2. Solving for R_2 gives

$$R_2 = \frac{-R_{\text{Tat}} + \sqrt{R_{\text{Tat}}^2 + 4C}}{2} \tag{2.30}$$

with

$$C = \frac{3R_3^2 \left(z_{\text{phos}}^0 - \langle h(r) \rangle \right)}{z_{\text{phos}}^0 - z_{\text{phos}}} - R_{\text{Tat}}^2. \tag{2.31}$$

2.4 Results

2.4.1 Bending and Bulk Modulus

Figure 2.2 shows the X-ray scattering intensity pattern from DOPC/DOPE (1:1) with Tat mole fraction $x_{\text{Tat}} = 0.034$. The diffuse lobes are due to equilibrium fluctuations that occur in these fully hydrated, oriented lipid/peptide samples. The intensity $I(\mathbf{q})$ in the diffuse patterns provide the absolute values of the form factors $F(q_z)$, which are the Fourier transforms of the electron density profile, through the relation $I(\mathbf{q}) = S(\mathbf{q})|F(q_z)|^2/q_z$, where $\mathbf{q} = (q_r, q_z)$, $S(q)$ is the structure interference factor, and q_z^{-1} is the usual LAXS approximation to the Lorentz factor [39, 65, 66]. The first step in the analysis takes advantage of the q_r dependence of the scattering to obtain the bending modulus K_c with results shown in Fig. 2.8. As positively charged Tat concentration was increased, the lamellar repeat spacing D generally increased in neutral lipid bilayers and decreased in negatively charged bilayers, consistent with changes in electrostatic repulsive interactions. With few exceptions, the water space between bilayers exceeded 20 Å.

2.4.2 Form Factors

From the K_c and B values obtained via the diffuse scattering analysis, the structure factor $S(\mathbf{q})$ is calculated, which leads to the absolute form factors $|F(q_z)| = I(\mathbf{q})/S(\mathbf{q})$. To estimate uncertainties on $|F(q_z)|$, we analyzed multiple diffuse scattering data obtained by sampling different lateral positions for each sample, which gave multiple form factors for a given sample. These form factors were averaged to give the average form factors and standard deviations for that sample (Fig. 2.9). Due to a small number of data sets for each sample, these standard deviations were noisy, so they were smoothed over adjacent 20 points. Average absolute form factors for five different membrane mimics are shown in Figs. 2.10, 2.11, 2.12, and 2.13. Vertical dashed lines indicate the "zero" position between

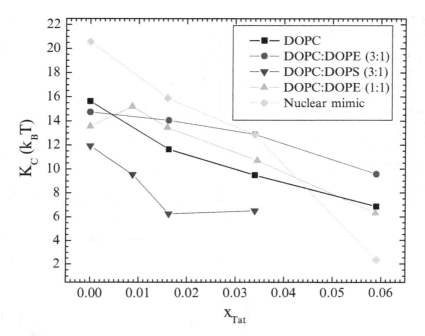

Fig. 2.8 Bilayer bending modulus, K_c, vs. Tat mole fraction x_{Tat}. D-spacings for DOPC:Tat mixtures varied from 64 to 68 Å, for DOPC:DOPE:Tat mixtures from 64 to 69 Å, for DOPC:DOPS:Tat (3:1) mixtures from 57 Å to > 100 Å (pure DOPS was unbound), and for nuclear mimic:Tat mixtures from unbound (nuclear mimic) to 64 Å. Estimated uncertainty in all values is $\sim \pm 2$

the lobes of diffuse data where $F(q_z)$ change sign. In almost all samples, the zero positions shift to larger q_z as Tat mole fraction increased, indicating a thinning of the membranes. The thinning effect will be quantified by fitting experimental form factors to models as will be discussed in Sect. 2.4.4.

2.4.3 Volume

Experimental and simulated volumes are given in Table 2.5. The simulated volume was obtained using the volume app [67] in the SIMtoEXP program [59]. The experimental Tat volume was calculated from the measured density assuming that the lipid volume was the same as with no Tat. In general, there may be an interaction volume between the peptide and the lipid membrane as previously reported for bacteriorhodopsin [68]. As lipid was present in excess to Tat, the partial molecular volume of the lipid is the same as with no Tat, so this way of calculating includes all the interaction volume in V_{Tat}. Comparison of V_{Tat} in water with the result for 5:1 Lipid:Tat suggests that the interaction volume may be negative, consistent with a net

Fig. 2.9 Three form factors obtained for three different lateral positions on a DOPC/Tat mixture with Tat mole fraction $x_{Tat} = 0.016$ (*top*). Standard deviations calculated by averaging these form factors are shown by *black solid line* (*bottom*). Because of the small number of data sets, the uncertainties are noisy, so for a model fitting purpose, they were smoothed over adjacent 20 points

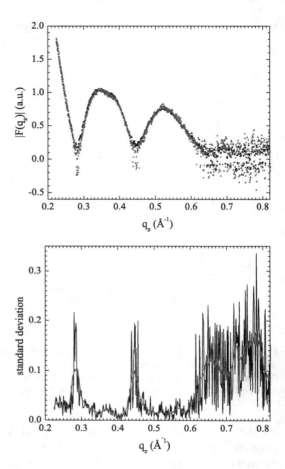

attractive interaction with lipid. Understandably, values of V_{Tat} were unreliable for small mole ratios of Tat:Lipid. Therefore we used simple additivity for those mimics not shown in Table 2.5 for the volumes used in the electron density profile modeling. All volumes obtained from the Gromacs MD simulations were somewhat smaller than the measured volumes, but it supports the Tat volume being closer to $1877\,\text{Å}^3$ than the outlying values obtained experimentally at small Tat concentrations. The measured volume is in good agreement with the value calculated from a peptide calculator website [69], which gave $1888\,\text{Å}^3$.

2.4.4 Electron Density Profile Modeling

We fit our measured X-ray form factors to the Tat-in-headgroup (THG) model described in Sect. 2.2.6. In all fits, the positions of component groups were free

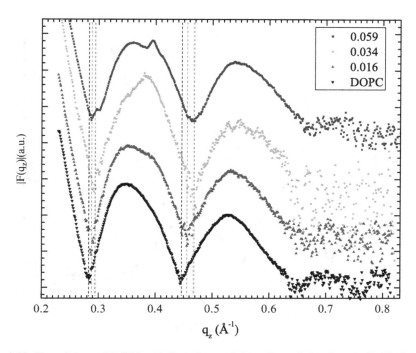

Fig. 2.10 Form factors of DOPC with Tat mixtures (arbitrarily scaled and vertically displaced) with increasing Tat mole fractions x_{Tat} indicated on figure legend. *Dashed vertical lines* roughly indicate the q_z values where the form factors are equal to zero between the lobes of diffuse data. Error bars are omitted for clarity

parameters, but we assumed that the lipid headgroup is somewhat rigid so that it cannot compress or expand. This assumption led to fixing the distance $z_{PC} - z_{CG}$ between the PC and CG components as well as the distance $z_{CG} - z_{HC}$ between the CG component and the Gibbs dividing surface for the hydrocarbon chains. We also constrained the width of Tat Gaussian σ_{Tat}. We fitted with three different values of widths, 2.5, 3.0, and 3.5, to study the range of variation due to the Tat width. We eventually constrained the Tat width because it tended to become unphysically small when it was set free. Without higher q_z data points, a very narrow feature in an electron density profile, which results in large form factors at high q_z, are not penalized.

Table 2.6 shows the model parameters that produced the best fits for DOPC with Tat. At lower Tat concentrations ($x_{Tat} = 0.016$ and 0.034), a smaller χ^2 value was obtained for smaller σ_{Tat}, consistent with its tendency to become unphysically small as noted in the previous paragraph. The widths of the headgroups σ_{PC} and σ_{CG} decreased from those of pure DOPC when Tat was added. It is also seen from Table 2.6 that the area per lipid A_L increased as the Tat concentration was increased from 0 to 0.034. An increase in A_L implies thinning of a bilayer because a lipid bilayer can be approximated as an incompressible fluid membrane. Another

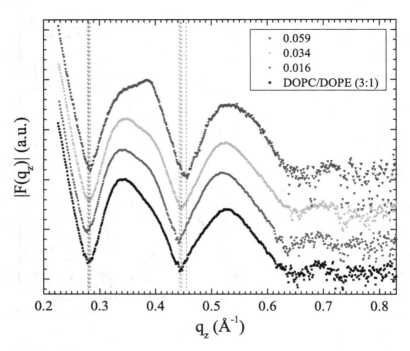

Fig. 2.11 Form factors of DOPC:DOPE (3:1) with Tat mixtures. The rest of the caption is the same as in Fig. 2.10

observed trend was that z_{Tat} increased as x_{Tat} was increased. Figure 2.14 shows the best fits and corresponding electron density profiles for DOPC with Tat.

As shown in Fig. 2.14, the membrane thickness can be defined as the distance D_{PP} between the PC components in the opposing leaflets or the distance D_{HH} between the maxima in the opposing leaflets. D_{HH} is more reliable than D_{PP} because it is a property of the total electron density of a bilayer and, therefore, does not depend strongly on the specific model employed for fitting the data. This point is illustrated in Fig. 2.15, which compares total electron density profiles resulting from best fits with three different Tat widths σ_{Tat}. While positions of Tat were sensitive to values of σ_{Tat}, the total electron density profiles were almost independent of σ_{Tat}. Essentially, other components, namely headgroups, adjusted their widths and positions so that the total electron density profile was about the same. In other words, the model was over parameterized. While the precise values of each parameter was less trustworthy, the total electron density profiles plotted in Fig. 2.15, when Fourier transformed, reproduced the experimental form factors very well and therefore are robust.

In contrast to D_{HH}, D_{PP} is a property that depends on lipid components, which are influenced by how the lipid is parsed (see Sect. 2.2.6) and what assumptions and constraints go into the specific model. A disadvantage of using D_{HH} as a measure of the membrane thickness is that D_{HH} is influenced by the electron density of Tat

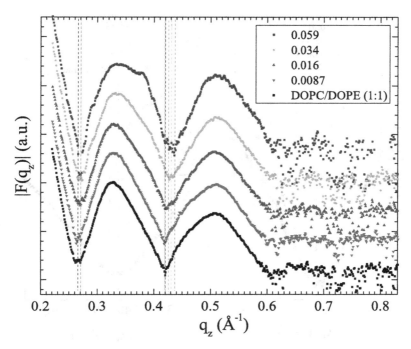

Fig. 2.12 Form factors of DOPC:DOPE (1:1) with Tat mixtures. The rest of the caption is the same as in Fig. 2.10

because the total electron density profile includes a contribution from the electron density of Tat. Especially when the mole fraction of Tat in a system becomes large, the Tat electron density contributes significantly to the total electron density profile. If Tat resided slightly outside of the PC component, the apparent membrane thickness measured by D_{HH} would be larger than D_{PP}. Then, even if the actual bilayer thickness defined by D_{PP} were reduced by the presence of Tat, the effect of thinning might not be obvious.

As described in the previous paragraph, the model parameters were sensitive to specific constraints and assumptions on the model, and as Fig. 2.15 shows, the position of Tat depended on σ_{Tat}. On the other hand, the total electron density profiles were seen to be less sensitive. Figure 2.16 compares the total electron density profiles at different Tat concentrations. Consistent with the form factors shifting to larger q_z as x_{Tat} increased, D_{HH} decreased as x_{Tat} increased. As argued earlier, a decrease in D_{HH} does not necessarily indicate a decrease in the bilayer thickness, and it could instead be attributed to deeper insertion of Tat into the bilayer. However, compared to the profile of DOPC alone, all three profiles with Tat deviate from the electron density of water at smaller $|z|$ when approached from the water region. This is illustrated in Fig. 2.17 that plots the difference between the total electron density profile of DOPC and those of DOPC with Tat. Negative values of

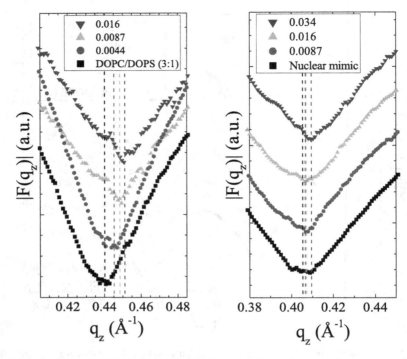

Fig. 2.13 Form factors of DOPC:DOPS (3:1) (*left*) and nuclear membrane mimic (*right*) with Tat mixtures. Portions of the form factors $|F(q_z)|$ that were not significantly distorted by mosaic spread scattering are shown. The most of the caption is the same as in Fig. 2.10

Table 2.5 Volume results at 37 °C

Experiments			
Tat in:	V_{lipid} (Å^3)	x_{Tat}	V_{Tat} (Å^3)
water			1877
DOPC:DOPE (3:1)	1288	0.167	1822
DOPC	1314	0.0246	676
DOPC:DOPS (3:1)	1298	0.0246	2613
Simulations			
Tat in:	V_{lipid} (Å^3)	Lipid:Tat	V_{Tat} (Å^3)
DOPC	1283	128:2	1694
DOPC	1294	128.4	1699

$\Delta\rho = \rho_{DOPC+Tat} - \rho_{DOPC}$ (the region labeled $\Delta\rho < 0$ in Fig. 2.17) indicate that the headgroup, which has excess electron density relative to water, shifted toward the bilayer center as Tat was added to the system, which implies bilayer thinning. This effect can also be seen in Fig. 2.16.

Fitting results for DOPC:DOPE (3:1) and DOPC:DOPE (1:1) are summarized in Tables 2.7 and 2.8, respectively, and the best fits and corresponding electron density

Table 2.6 Fitting results for DOPC membranes for the THG (Tat in headgroup) model. $z_{PC} - z_{CG} = 3.1$ Å and $z_{CG} - z_{HC} = 1.3$ Å in all fits. Units of all symbols are Å except for χ^2 (unitless) and A_L (Å^2)

x_{Tat}	0	0.016	0.016	0.016	0.034	0.034	0.034	0.059	0.059	0.059
χ^2	2961	1554	1570	1581	1563	1587	1607	2342	2338	2363
z_{PC}	18.1	18.0	17.9	17.9	17.8	17.7	17.6	17.8	17.8	17.7
σ_{PC}	2.52	2.14	2.17	2.18	1.86	1.92	1.93	2.02	1.97	1.93
z_{CG}	15.0	14.9	14.8	14.8	14.7	14.6	14.5	14.7	14.7	14.6
σ_{CG}	3.00	2.62	2.64	2.66	2.22	2.30	2.31	2.58	2.27	2.14
z_{HC}	13.7	13.6	13.5	13.5	13.4	13.3	13.2	13.4	13.4	13.3
σ_{HC}	3.00	2.69	2.84	2.95	2.65	2.82	3.01	2.47	2.58	2.83
σ_{CH_3}	3.20	3.19	3.22	3.24	3.37	3.43	3.47	2.70	2.70	2.74
z_{Tat}	NA	12.9	13.4	14.2	13.1	13.8	14.4	15.2	15.2	15.7
σ_{Tat}	NA	2.5[a]	3.0[a]	3.5[a]	2.5[a]	3.0[a]	3.5[a]	2.5[a]	3.0[a]	3.5[a]
A_L	71.5	72.4	72.5	72.7	73.6	74.0	74.4	73.6	73.5	73.9

[a]Paramters were fixed

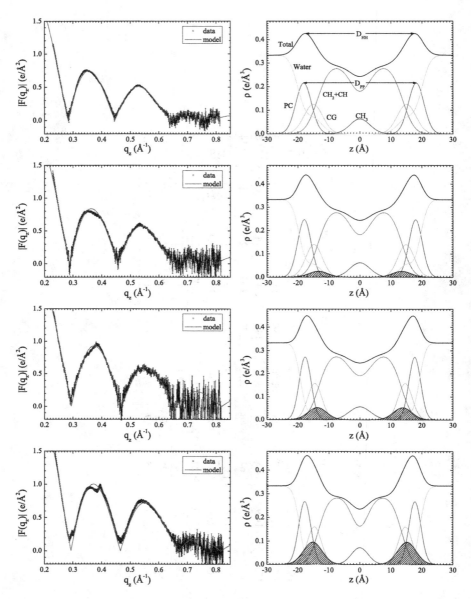

Fig. 2.14 Best fits to DOPC form factors (*left*) and the corresponding electron density profiles (*right*) with x_{Tat} = 0, 0.016, 0.034, and 0.059 (from *top* to *bottom*). The *solid lines* corresponding to varioius bilayer molecular components are labelled in the top electron density profile (EDP). The Tat EDP is a *solid black line* with diagonal line underfill

Fig. 2.15 Comparison of total electron density profiles corresponding to best fits using different Tat widths σ_{Tat}, 2.5, 3.0, and 3.5. The mole fraction of Tat x_{Tat} was 0.016 (*top*), 0.034 (*middle*), and 0.059 (*bottom*). While different values of σ_{Tat} resulted in different positions of Tat, the total electron density profiles were almost identical and independent of σ_{Tat}

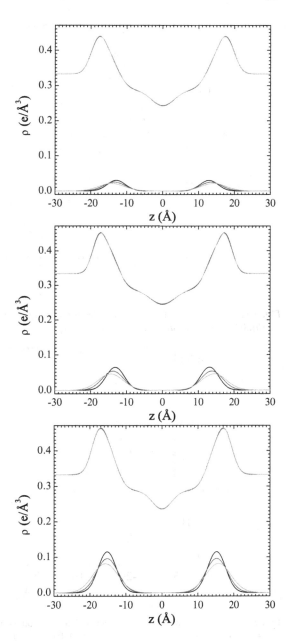

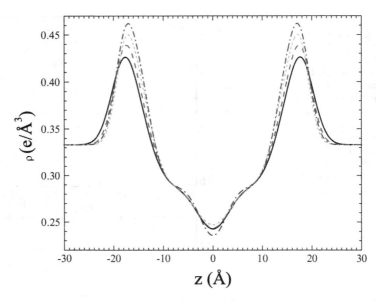

Fig. 2.16 Comparison of DOPC total electron density profiles at $x_{Tat} = 0$ (*black solid*), 0.016 (*dash*), 0.034 (*short dash*), and 0.059 (*dash dot*)

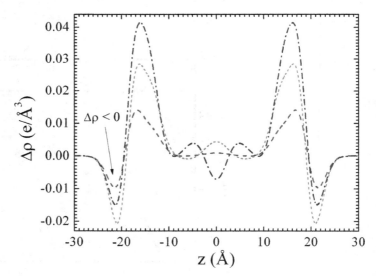

Fig. 2.17 Difference between total electron density profiles of DOPC with Tat and that of DOPC $\Delta\rho = \rho_{DOPC+Tat} - \rho_{DOPC}$. $x_{Tat} = 0.016$ (*dash*), 0.034 (*short dash*), and 0.059 (*dash dot*). Positive $\Delta\rho$ means excess electron density due to presence of Tat. The region labeled $\Delta\rho < 0$ indicates that the electron dense headgroup moved closer to the bilayer center upon addition of Tat, which is equivalent to bilayer thinning

Table 2.7 Fitting Results for DOPC:DOPE (3:1) membranes for the THG model. $z_{PC} - z_{CG} = 3.1$ Å and $z_{CG} - z_{HC} = 1.3$ Å in all fits. Units of all symbols are Å except for χ^2 (unitless) and A_L (Å²)

x_{Tat}	0	0.016	0.016	0.016	0.034	0.034	0.034	0.059	0.059	0.059
χ^2	924.5	4972	4985	4994	6758	6826	6863	2293	2280	2296
z_{PC}	18.3	18.5	18.5	18.4	18.5	18.4	18.3	18.2	18.2	18.1
σ_{PC}	2.66	2.23	2.26	2.27	2.25	2.31	2.34	2.31	2.19	2.11
z_{CG}	15.2	15.4	15.4	15.3	15.4	15.3	15.2	15.1	15.1	15.0
σ_{CG}	2.92	2.63	2.65	2.69	2.52	2.58	2.63	2.40	2.20	2.01
z_{HC}	13.9	14.1	14.1	14.0	14.1	14.0	13.9	13.8	13.8	13.7
σ_{HC}	2.73	2.70	2.83	2.91	2.86	2.79	2.84	2.25	2.38	2.60
σ_{CH_3}	3.24	2.94	2.97	2.98	2.87	2.90	2.91	2.63	2.61	2.65
z_{Tat}	NA	13.5	14.0	15.0	14.3	14.9	16.0	16.3	16.4	16.9
σ_{Tat}	NA	2.5	3.0	3.5	2.5	3.0	3.5	2.5	3.0	3.5
A_L	70.9	69.8	69.9	70.1	69.5	70.0	70.6	71.3	71.4	71.7

Table 2.8 Fitting Results for DOPC:DOPE (1:1) membranes for the THG model. $z_{PC} - z_{CG} = 3.1$ Å and $z_{CG} - z_{HC} = 1.3$ Å in all fits. Units of all symbols are Å except for χ^2 (unitless) and A_L (Å²)

x_{Tat}	0	0.016	0.016	0.016	0.034	0.034	0.034	0.059	0.059	0.059
χ^2	2961	1554	1570	1581	1563	1587	1607	2342	2338	2363
z_{PC}	18.1	18.0	17.9	17.9	17.8	17.7	17.6	17.8	17.8	17.7
σ_{PC}	2.52	2.14	2.17	2.18	1.86	1.92	1.93	2.02	1.97	1.93
z_{CG}	15.0	14.9	14.8	14.8	14.7	14.6	14.5	14.7	14.7	14.6
σ_{CG}	3.00	2.62	2.64	2.66	2.22	2.30	2.31	2.58	2.27	2.14
z_{HC}	13.7	13.6	13.5	13.5	13.4	13.3	13.2	13.4	13.4	13.3
σ_{HC}	3.00	2.69	2.84	2.95	2.65	2.82	3.01	2.47	2.58	2.83
σ_{CH_3}	3.20	3.19	3.22	3.24	3.37	3.43	3.47	2.70	2.70	2.74
z_{Tat}	NA	12.9	13.4	14.2	13.1	13.8	14.4	15.2	15.2	15.7
σ_{Tat}	NA	2.5	3.0	3.5	2.5	3.0	3.5	2.5	3.0	3.5
A_L	71.5	72.4	72.5	72.7	73.6	74.0	74.4	73.6	73.5	73.9

profiles are shown in Figs. 2.18 and 2.19 at the end of this subsection. Figure 2.20 plots total electron density profiles, showing increasing electron density in the headgroup region as Tat concentration increased, similarly to DOPC/Tat systems shown in Fig. 2.16.

Figure 2.21 summarizes the results for bilayer thickness as a function of Tat mole fraction x_{Tat}. In all cases, D_{HH} was smaller than D_{PP}, consistent with the results that the value of Tat position z_{Tat} was smaller than that of PC headgroup position z_{PC}. The CG headgroup also carries high average electron density and is located closer to the bilayer center than the PC headgroup. Therefore, in general, D_{HH} is smaller than D_{PP} even without Tat. Figure 2.22 compares Tat position to the PC headgroup position, reemphasizing the result that Tat is located inside the PC headgroup. We note, however, that D_{PP} in our models is the average PC-PC distance and not necessarily the same as local bilayer thickness near a Tat peptide. It is reasonable to expect that the perturbation of bilayer structure due to Tat is largest near Tat and decays as a function of lateral distance from Tat. In Sect. 2.4.5, we discuss local perturbation of a DOPC bilayer measured in MD simulations. Finally, Fig. 2.23 plots area per lipid as a function of Tat mole fraction. Consistent with bilayer thinning, area per lipid was found to increase in most cases. We could not obtain electron density profiles for DOPC:DOPS (3:1) and the nuclear membrane mimic, due to insufficient diffuse X-ray scattering by Tat charge neutralization of these negatively charged membranes, which rendered extraction of X-ray form factors unreliable.

We also studied how the goodness of fit varied as the position of the Tat Gaussian was varied. Figure 2.24 plots χ^2 as a function of the fixed Tat position z_{Tat}. We found that the two models, THG (Tat-in-headgroup region) and THC (Tat-in-hydrocabon-chain region), resulted in similar electron density profiles, yielding similar χ^2 values when Tat was placed near the hydrocarbon-water interface region. In the THC model, the error function representing the hydrocarbon chain region became wider as Tat was placed closer to the interface region such that the total density profile calculated from the THC model was very similar to that calculated from the THG model. In general, while the total electron density profile is well determined by our modeling procedures, the values of the parameters for the components are not as well determined as the agreement of the fit to the data may suggest. In many cases, we found multiple local minima in the fitting landscape, including one with Tat closer to the center of the bilayer as shown in Fig. 2.24. χ^2 calculated at these local minima tended to be smaller for larger concentration of Tat. We also found that χ^2 with z_{Tat} in the hydrocarbon chain region and headgroup region was almost equal for the largest value of x_{Tat} for DOPC:DOPE (1:1) bilayer. The MD simulations performed by Dr. Kun Huang suggested that the interior positions of Tat were artifacts of our model, at least for DOPC bilayers. The simulation results are found in Sect. 2.4.5.

As seen from Table 2.6, the widths of the headgroup components became smaller as Tat concentration increased. This decrease seemed somewhat unreasonable; if Tat causes a bilayer to locally become thinner, we would expect the headgroup components to become wider. Therefore, we also fitted a model with lower bounds on these headgroup widths. Namely, the minimum values of the widths of the headgroup components, PC and CG, were constrained to be greater than or equal

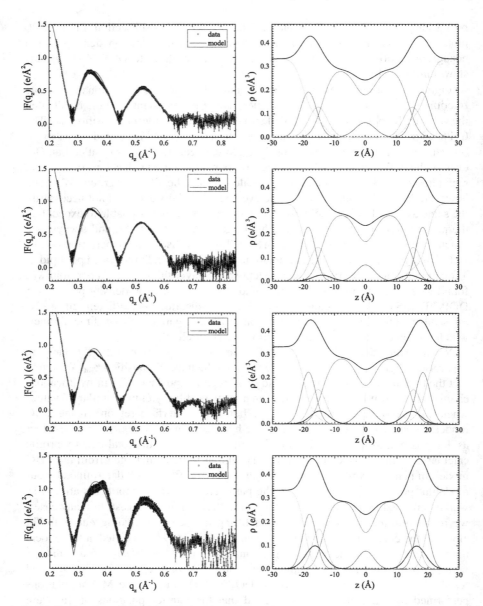

Fig. 2.18 Best fits to DOPC:DOPE (3:1) form factors (*left*) and the corresponding electron density profiles (*right*) with $x_{Tat} = 0$, 0.016, 0.034, and 0.059 (from *top* to *bottom*). The Tat EDP is a *solid black line* with diagonal line underfill

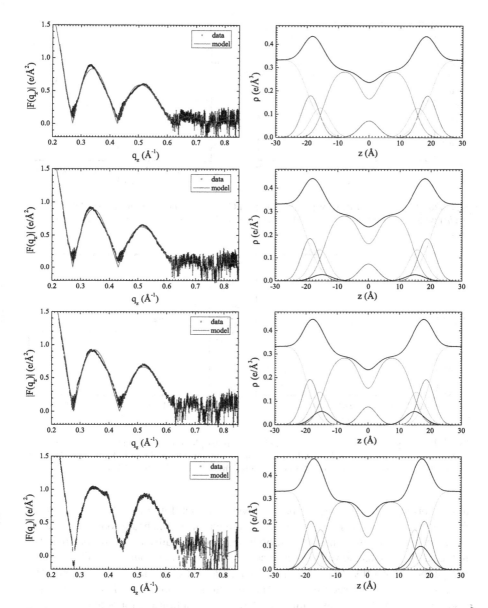

Fig. 2.19 Best fits to DOPC:DOPE (1:1) form factors (*left*) and the corresponding electron density profiles (*right*) with $x_{\text{Tat}} = 0$, 0.016, 0.034, and 0.059 (from *top* to *bottom*). The Tat EDP is a *solid black line* with diagonal line underfill

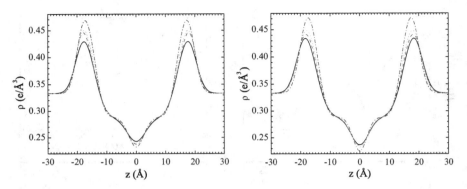

Fig. 2.20 Total electron density profiles for DOPC:DOPE (3:1) (*left*) and DOPC:DOPE (1:1) (*right*) with Tat mole fraction $x_{Tat} = 0$ (*solid*), 0.016 (*dash*), 0.034 (*short dash*), and 0.059 (*dash dot*)

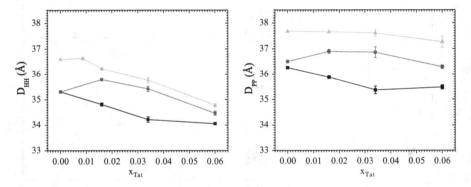

Fig. 2.21 Bilayer thickness, D_{HH} (*left*) and D_{PP} (*right*) plotted against Tat mole fraction x_{Tat}. *Squares* (DOPC), *circles* (DOPC:DOPE (3:1)), and *triangles* (DOPC:DOPE (1:1)). Error bars are standard deviations from imposing Tat Gaussian widths, $\sigma_{Tat} = 2.5$, 3.0 or 3.5 Å

to the corresponding values for pure bilayers without Tat. Table 2.9 shows results from fitting the data with lower bounds on the widths of the headgroup components for DOPC/Tat systems. In all cases, both headgroup widths, σ_{PC} and σ_{CG}, resulted in the same value as the value of their corresponding lower bounds. Similarly to fits with unbound widths, $D_{PP} = 2z_{PC}$ decreased as Tat concentration increased. The biggest difference between these bound fits and the unbound fits is in Tat position z_{Tat}. Figure 2.25 plots z_{Tat} as a function of Tat mole fraction x_{Tat} for both fits with and without lower bounds. While z_{Tat} increased as x_{Tat} increased for fits without bounds, z_{Tat} stayed more or less constant for fits with the bounds. Moreover, Table 2.9 shows that Tat was located closer to the PC headgroup than the CG headgroup for fits with the lower bounds. Thus, depth of Tat insertion was influenced strongly by the lipid headgroup widths. In order to gain better understanding of location of Tat in DOPC bilayers, we now turn to MD simulations.

Table 2.9 Fitting Results for the THG model with the lower bounds on the headgroup widths for DOPC membranes. $z_{PC} - z_{CG} = 3.1$ and $z_{CG} - z_{HC} = 1.3$ in all fits. Units of all symbols are Å except for χ^2 (unitless) and A_L (Å^2)

x_{Tat}	0	0.016	0.016	0.016	0.034	0.034	0.034	0.059	0.059	0.059
χ^2	2961	1853	1979	2118	2398	2893	3414	3160	4298	5539
z_{PC}	18.1	17.8	17.8	17.8	17.4	17.4	17.4	17.5	17.4	17.3
σ_{PC}	2.5	2.5[a]	2.5[a]	2.5[a]	2.5[a]	2.5[a]	2.5[a]	2.5[a]	2.5[a]	2.5[a]
z_{CG}	15.0	14.7	14.7	14.7	14.3	14.3	14.3	14.4	14.4	14.3
σ_{CG}	3.0	3.0[a]	3.0[a]	3.0[a]	3.0[a]	3.0[a]	3.0[a]	3.0[a]	3.0[a]	3.0[a]
z_{HC}	13.7	13.4	13.4	13.4	13.0	13.0	13.0	13.1	13.0	12.9
σ_{HC}	3.0	2.7	2.7	2.7	2.7	2.7	2.7	2.7	2.7	2.7
σ_{CH_3}	3.2	3.1	3.1	3.1	3.6	3.6	3.7	2.6	2.6	2.5
z_{Tat}	–	16.9	16.8	17.0	16.4	16.5	16.7	16.3	16.6	17.1
σ_{Tat}	–	2.5	3.0	3.5	2.5	3.0	3.5	2.5	3.0	3.5
A_L	71.5	73.5	73.5	73.5	75.6	75.6	75.6	75.0	75.4	75.9

[a]Parameters with a lower bound as described in the text

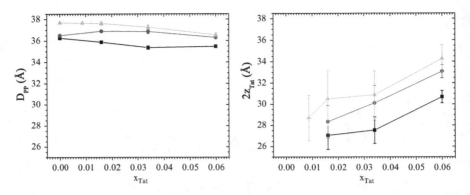

Fig. 2.22 Bilayer thickness, D_{PP} (*left*) and twice Tat position $2z_{Tat}$ (*right*) plotted against Tat mole fraction x_{Tat}. *Squares* (DOPC), *circles* (DOPC:DOPE (3:1)), and *triangles* (DOPC:DOPE (1:1)). Error bars are standard deviations from imposing Tat Gaussian widths, $\sigma_{Tat} = 2.5, 3.0$ or 3.5 Å. The data points of D_{PP} (left) are identical to those in Fig. 2.21, but the left axis is adjusted to facilitate comparison against $2z_{Tat}$

Fig. 2.23 Area per lipid plotted against Tat mole fraction x_{Tat}. *Squares* (DOPC), *circles* (DOPC:DOPE (3:1)), and *triangles* (DOPC:DOPE (1:1)). Error bars are standard deviations from imposing Tat Gaussian widths, $\sigma_{Tat} = 2.5,$ 3.0 or 3.5 Å

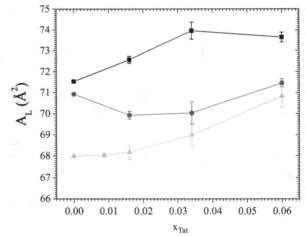

2.4.5 Molecular Dynamics Simulations

Due to slow relaxation in lipid bilayers and limited force field accuracy, good agreement may be difficult to reach between experimental and MD simulation calculated form factors. Consequently, we carried out several constrained simulations at various A_L and z_{Tat} as described in Sect. 2.2.7. We then compared the simulated and experimental form factors $F(q_z)$. Figure 2.26 compares simulated and experimental DOPC form factors. The simulated form factor shifted to larger q_z as the area per lipid increased, consistent with results in Sect. 2.4.4. We determined that the simulation at $A_L = 70$ Å^2 best reproduced the experimental form factor, yielding

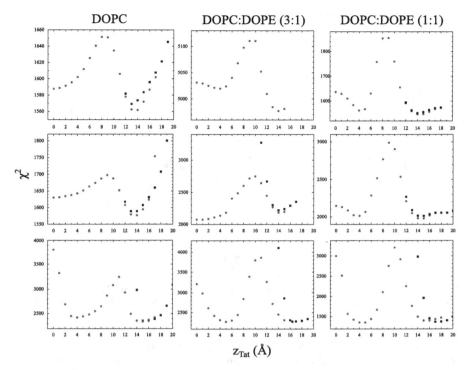

Fig. 2.24 χ^2 as a function of z_{Tat} for DOPC (*left column*), DOPC:DOPE (3:1) (*middle*), and DOPC:DOPE (1:1) (*right*) with x_{Tat} = 0.016 (*top row*), 0.034 (*middle*), and 0.059 (*bottom*). σ_{Tat} = 3.0. The THG model (*squares*) and the THC model (*circles*)

Fig. 2.25 z_{Tat} as a function of Tat mole fraction x_{Tat} for fits with lower bounds on the headgroup widths (*circles*) and without lower bounds (*squares*). Error bars are standard deviations from imposing Tat Gaussian widths, σ_{Tat} = 2.5, 3.0 or 3.5 Å

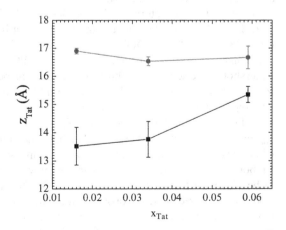

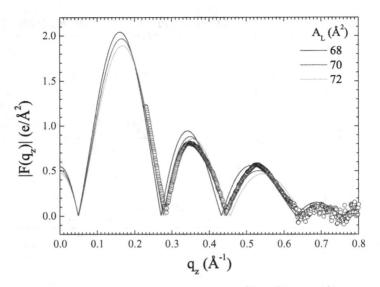

Fig. 2.26 MD simulated form factors for DOPC at $A_L = 68\,\text{Å}^2$, $70\,\text{Å}^2$, and $72\,\text{Å}^2$ compared to the experimental form factor (*open circles*) scaled vertically to best match the form factor for $70\,\text{Å}^2$

the smallest χ^2 value. However, the simulated form factor for $A_L = 72\,\text{Å}^2$ best matched the experimental form factor near $q_z = 0.3\,\text{Å}^{-1}$, which suggests that a better match might lie between 70 and $72\,\text{Å}^2$. This case was not investigated further. The electron density profile from the best matching simulation is shown in Fig. 2.27 with atoms in the simulation parsed into the same molecular component groups as in the model used in Sect. 2.4.4.

Simulated form factors $|F^{\text{sim}}|$ (see Sect. 2.3.1) for DOPC:Tat (2 Tat molecules in 128 DOPC molecules), where there is one Tat in each monolayer, are shown in Fig. 2.28, and $|F^{\text{sim}}|$ for DOPC:Tat (4 Tat in 128 DOPC) are shown in Fig. 2.29 for z_{Tat} constrained to 18, 16, and 14 Å. For DOPC:Tat (128:2), $|F^{\text{sim}}|$ overshot and undershot in the second and third lobe regions, respectively. For DOPC:Tat (128:4), $|F^{\text{sim}}|$ agreed well with $|F^{\text{exp}}|$ in the second lobe region but undershot in the third lobe region. Quantitative comparison of simulated form factors to the experimental form factor is shown in Table 2.10. We found the best match at $A_L = 72\,\text{Å}^2$ and $z_{\text{Tat}} = 18\,\text{Å}$ for DOPC:Tat (128:2). The best match for DOPC:Tat (128:4) was found when Tats were constrained at 18 Å away from the bilayer center with $A_L = 76\,\text{Å}^2$. At both Tat concentrations, the agreement worsened when Tat was constrained to be closer to the center of the bilayer. When Tats were constrained to be 5 Å from the bilayer center, we observed a formation of water pores in the simulation. However, as shown in Fig. 2.30, the corresponding simulated form factor did not agree well with the experimental form factor. Thus, comparison of the experimental and simulated form factors indicates that Tat is located in the headgroup position; Tat is not located in the hydrocarbon region.

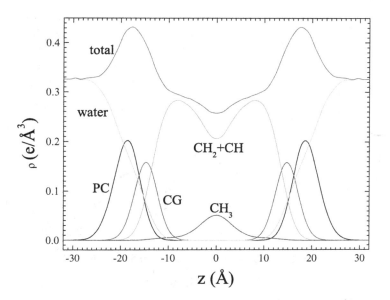

Fig. 2.27 Simulated, symmetrized electron density profile for DOPC at $A_L = 70 \text{ Å}^2$ as a function of the distance from the bilayer center. Each component profile is labeled with its name: PC (phosphate-choline), CG (carbonyl-glycerol), CH_2+CH (methylene-methine combination), CH_3 (terminal methyl). The sum of all the components is labeled as total

Figure 2.31 plots the bilayer thickness defined as D_{HH} and D_{PP}, area per lipid A_L, and Tat position z_{Tat} with results from the modeling approach in Sect. 2.4.4. Consistent with the modeling, simulation indicated decreasing bilayer thickness and increasing area per lipid as Tat concentration increased. Tat is found in the headgroup position in both simulations and modeling, but z_{Tat} is consistently larger in simulations.

We also obtained local phosphorus-phosphorus distance D_{phos} (see Fig. 2.32) shown in Table 2.11 using the methods described in Sect. 2.3.2. For comparison, the average bilayer thickness defined by the average phosphorus-phosphorus distance $\langle D_{phos} \rangle$ is also shown in Table 2.11 for simulations with $z_{Tat} = 16 \text{ Å}$ and 18 Å. $\langle D_{phos} \rangle$ was measured using the electron density profile of phosphorus atoms. The D_{phos} column shows that the membrane thickness was smaller near Tats as compared to the average thickness given in the $\langle D_{phos} \rangle$ column. A decrease in the local membrane thickness $\Delta t = \langle D_{phos}^0 \rangle - D_{phos}$ with respect to the average thickness $\langle D_{phos}^0 \rangle = 36.3 \text{ Å}$ of a pure DOPC bilayer was larger for higher concentrations of Tat.

Assuming that the two leaflets are decoupled, we also estimated the position of phosphorus atoms z_{phos} near Tats using $z_{phos} = D_{phos} - \langle D_{phos}^0 \rangle / 2$, shown in the z_{phos} column in Table 2.11. The z_{phos} values are smaller than $D_{phos}/2$ because it is assumed that lipids in the other leaflet are unperturbed, having thickness $= \langle D_{phos}^0 \rangle / 2$. The calculation of z_{phos} assumes that Tat in different leaflets do not

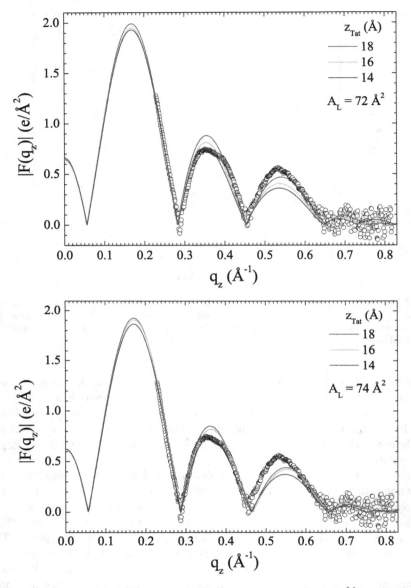

Fig. 2.28 MD simulated form factors for DOPC with $x_{Tat} = 0.015$ at $A_L = 72\,\text{Å}^2$ (*top*) and $74\,\text{Å}^2$ (*bottom*), with $z_{Tat} = 18\,\text{Å}$, $16\,\text{Å}$, and $14\,\text{Å}$ compared to the experimental form factor (*open circles*) scaled vertically to best match the form factor for $z_{Tat} = 18\,\text{Å}$

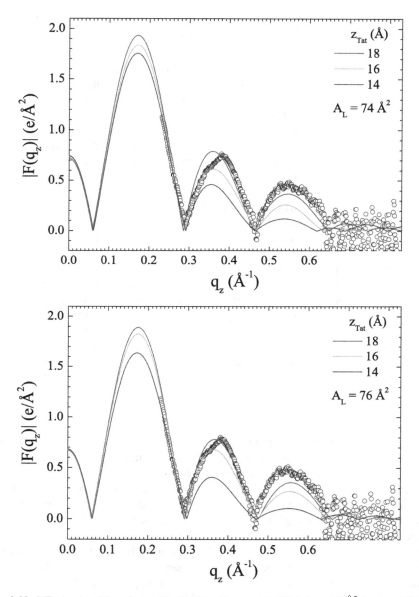

Fig. 2.29 MD simulated form factors for DOPC with $x_{Tat} = 0.030$ at $A_L = 74$ Å^2 (*top*) and 76 Å^2 (*bottom*), with $z_{Tat} = 18$ Å, 16 Å, and 14 Å compared to the experimental form factor (*open circles*) scaled vertically to best match the form factor for $z_{Tat} = 18$ Å

Table 2.10 Comparison of the simulated form factors to the experimental form factors. a is an overall scaling factor described in Sect. 2.3.1.

$x_{Tat} = 0.015$				$x_{Tat} = 0.030$			
A_L (Å^2)	z_{Tat} (Å)	a	χ^2	A_L (Å^2)	z_{Tat} (Å)	a	χ^2
70	18	0.621	60.1	72	18	0.596	49
70	16	0.568	69.1	72	16	0.476	82
70	14	0.439	131	72	14	0.307	248
70	12	0.285	391	72	12	0.153	607
70	10	0.199	440	72	10	0.196	78
70	8	0.196	374	72	8	0.114	275
70	5	0.159	527	72	5	0.095	438
72	18	0.72	18.0	74	18	0.617	24
72	16	0.65	24.9	74	16	0.514	40
72	14	0.6	31.4	74	14	0.394	135
72	12	0.426	104	74	12	0.147	1092
72	10	0.219	443	74	10	0.125	334
72	8	0.205	336	74	8	0.101	496
72	5	0.165	448	74	5	0.129	424
74	18	0.722	21.3	76	18	0.648	15
74	16	0.704	25.9	76	16	0.573	30
74	14	0.631	25.7	76	14	0.376	158
74	12	0.412	81.9	76	12	0.172	1072
74	10	0.312	194	76	10	0.147	504
74	8	0.246	351	76	8	0.098	535
74	5	0.177	427	76	5	0.139	183

overlap in the plane, which might not be reasonable at the higher concentration. Therefore, smaller values of z_{phos} at higher Tat concentration may partly be due to the bilayer compressing both from above and below.

Table 2.11 also lists the position of guanidinium groups averaged over all arginines in the z_{guan} column. z_{guan} was obtained from the peak position of the electron density profile of the guanidinium groups as shown in Fig. 2.33. As Fig. 2.33 shows, the distribution of the guanidinium groups was broad and asymmetric with its peak at smaller z than the center of the distribution, indicating that more arginines are located closer to the hydrocarbon region than to the water. This is in contrast with amine groups in lysines whose distribution was peaked in the water region as shown by the blue curve in Fig. 2.33. Table 2.11 shows that $z_{guan} > z_{phos}$ but $z_{guan} < \langle D_{phos} \rangle / 2$, indicating that the guanidinium group can be considered inside or outside of the phosphorus atoms depending on whether local z_{phos} or average thickness $\langle D_{phos} \rangle / 2$ is considered.

Table 2.12 shows the Tat perturbation lateral decay length R_2 estimated using the method described in Sect. 2.3.3. The H_{Tat} column lists the FWHM values of the Tat electron density profile. The R_{Tat} column was calculated by assuming that

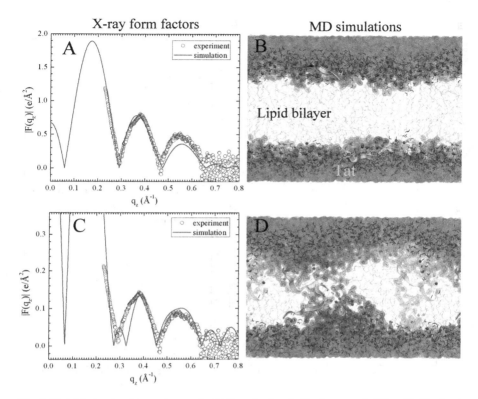

Fig. 2.30 MD simulated form factors (*solid lines* in *A* and *C*) of $x_{Tat} = 0.030$, with Tat fixed at $z_{Tat} = 18$ Å (*panel A*) and 5 Å (*panel C*) from the bilayer center compared to experimental form factors (*open circles*) scaled vertically to best fit the simulated form factors. Corresponding snapshots are shown in *Panels B* and *D* in which the lipid chains are represented as *grey sticks* on a *white background*, Tats are ribbons, phosphate groups are circles, and water is the rest

Tat is a cylinder with radius R_{Tat} and height H_{Tat} with the experimentally measured Tat volume. A cylindrical shape was chosen to reflect the azimuthal symmetry of the fluid bilayer. We did not consider other rotationally symmetric shapes. The calculation of R_2 involves an assumption that Tats in different leaflets do not overlap in the xy-plane, which might not be justified at the higher concentration. Therefore, values of R_2 for $x_{Tat} = 0.030$ are omitted in the table.

The χ^2 values obtained by comparing $|F^{sim}|$ to $|F^{exp}|$ in Table 2.10 indicated that Tat lay between the simulated values of 16 Å and 18 Å, and A_L lay between the simulated values of 72 Å^2 and 74 Å^2, so averages were obtained from these four combinations of z_{Tat} and A_L, weighted inversely with their χ^2. Figure 2.34 summarizes Tat's effect on a DOPC bilayer, based on the weighted average values shown in Table 2.13. Tat is modeled as a cylinder with height $H_{Tat} = 8.7$ Å and

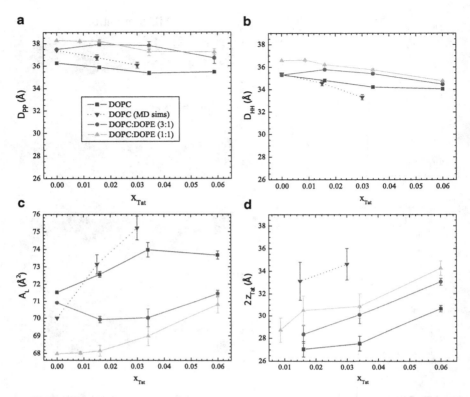

Fig. 2.31 (**a**) Bilayer thickness, D_{PP}; (**b**) Bilayer thickness, D_{HH}; (**c**) Area/lipid, A_L; (**d**) Twice the Tat location, $2z_{Tat}$: all plotted vs. Tat mole fraction x_{Tat}. Error bars for the experimental data points are standard deviations from imposing Tat Gaussian widths, $\sigma = 2.5$, 3.0, or 3.5 Å. Inverted *triangles* connected with a *dotted line* are results from MD simulations, averaging the values from simulations with the four smallest χ^2 for each x_{Tat} in Table 2.10, each weighted by $1/\chi^2$, with standard deviations shown. Samples are listed in the legend in panel A

radius $R_{Tat} = 8.3$ Å centered at $z_{Tat} = 17.1$ Å. The phosphorus atoms within the suppressed region (see Sect. 2.3.3) are positioned at $z_{phos} = 14.6$ Å. Assuming a simple linear ramp in z_{phos}, Fig. 2.34 indicates a ring of boundary lipids that extends twice as far in R as Tat itself. Although the guanidinium electron density profile was broad (Fig. 2.33), indicating that some were pointing away from the bilayer relative to the center of Tat, more were pointing towards the bilayer center as indicated in Fig. 2.34.

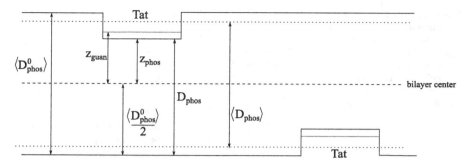

Fig. 2.32 Definitions of quantities relevant to Table 2.11. The *black solid line* is the profile of the phosphorus position. The *solid line* indicates the peak position of the guanidinium distribution. *Dotted lines* are the phosphorus position averaged over all lipids in each monolayer. In this picture, Tat is assumed to influence lipids only in the monolayer it binds

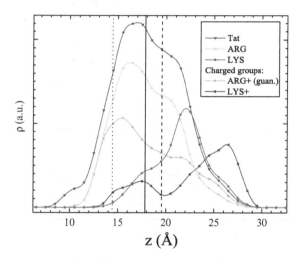

Fig. 2.33 Electron density profiles of Tat, arginine, lysine, guanidinium groups, and amine groups for DOPC:Tat (128:2) at $A_{\mathrm{L}} = 72\,\text{Å}^2$ and $z_{\mathrm{Tat}} = 18\,\text{Å}$. The *black solid line* indicates the phosphorus atom position for the pure DOPC bilayer, the *dashed line* the choline group, and *dotted line* the carbonyl-glycerol group. z_{guan} was obtained from the peak position of the electron density profile of the guanidinium groups. Curves are arbitrarily scaled in the vertical direction

2.5 Discussion

Given that 8 of the 11 amino acids in Tat are arginines and lysines, one would have suggested 20 years ago that highly charged Tat would partition strongly into solution rather than being associated with lipid bilayers. By contrast, but in agreement with more recent perspectives on arginine partitioning into the interfacial region [70], we find that Tat interacts with lipid bilayers, even with neutral DOPC

Table 2.11 Local bilayer structural quantities obtained at various constrained A_L and z_{Tat} at different Tat mole fraction x_{Tat}. Units of all symbols are Å except for x_{Tat} (unitless), χ^2 (unitless), and A_L (Å^2). $\langle D_{phos}^0 \rangle/2 = 18.2\,\text{Å}$

x_{Tat}	A_L	z_{Tat}	$\langle D_{phos} \rangle/2$	$D_{phos}/2$	Δt	z_{phos}	z_{guan}	χ^2
0.015	72	18	17.8	16.4	3.5	14.7	15.5	18.0
0.015	72	16	18.1	16.5	3.3	14.9	14.5	24.9
0.015	74	18	17.5	16.5	3.3	14.9	16.5	21.3
0.015	74	16	17.5	16.1	4.2	14.0	13.5	25.9
0.030	74	18	17.7	16.3	3.7	14.5	15.5	24.3
0.030	74	16	17.7	15.6	5.1	13.1	13.5	40.1
0.030	76	18	17.1	16.0	4.3	13.9	16.5	14.8
0.030	76	16	17.5	15.7	4.9	13.3	14.5	30.4

Table 2.12 Lateral decay length of Tat perturbation. The simulation box size $R_3 = 38\,\text{Å}$ (see Sect. 2.3.3)

x_{Tat}	A_L (Å^2)	z_{Tat} (Å)	H_{Tat} (Å)	R_{Tat} (Å)	R_2 (Å)
0.015	72	18	9.2	8.1	15.0
0.015	72	16	9.4	8.0	9.0
0.015	74	18	8.6	8.3	23.9
0.015	74	16	7.6	8.9	20.4
0.030	74	18	7.6	8.9	NA
0.030	74	16	7.7	8.8	NA
0.030	76	18	7.6	8.9	NA
0.030	76	16	7.8	8.7	NA

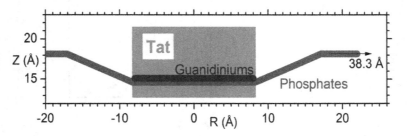

Fig. 2.34 Location of Tat in DOPC bilayer. Tat is represented as a cylinder, z is the distance from the bilayer center, and R is the in-plane distance from the center of Tat. The average z of the lipid phosphates as a function of R and the arginine guanidiniums are shown

and DOPC:DOPE mixtures, as well as with negatively charged DOPC:DOPS and nuclear membrane mimic lipid mixtures. This paper presents multiple lines of evidence for a Tat/membrane interaction. Figure 2.8 shows that Tat decreases the bending modulus. Although one could argue that such a decrease is only apparent and could instead be due to local changes in membrane spontaneous curvature [71], either interpretation supports a Tat-bilayer interaction. The changes with increasing Tat concentration in the X-ray membrane form factors in Figs. 2.10, 2.11, 2.12, and 2.13 shows that Tat affects membrane structure, and the shift of the zero positions to higher q_z suggests thinning. Thinning is substantiated by quantitative analysis of the X-ray data and by MD simulations. Figure 2.31a shows that the average membrane

Table 2.13 Weighted average quantities. Units of all symbols are Å except for x_{Tat} (unitless) and A_L (Å^2)

x_{Tat}	A_L	z_{Tat}	$\langle D_{phos} \rangle$	D_{phos}	Δt	H_{Tat}	R_{Tat}	R_2	z_{phos}	z_{guan}
0.015	72.9	17.1	35.4	32.7	3.6	8.7	8.3	17.1	14.6	15.1
0.030	75.2	17.3	34.8	31.9	4.4	7.7	8.8	NA	13.8	15.4

thickness, as measured by the distance D_{PP} between the phosphate-choline groups on opposite surfaces, decreases with increasing Tat concentration. Similar thinning is shown in Fig. 2.31b for the distance D_{HH} between the maxima in the electron density profiles of opposite surfaces. Compared to D_{PP}, D_{HH} is pulled towards both the carbonyl/glycerol groups and Tat because both have electron densities ($\sim$0.4 e/Å^3) greater than water ($\sim$0.33 e/Å^3) or hydrocarbon ($\sim$0.3 e/Å^3). Although the thinning shown in Figs. 2.31a, b is not large, it obviously requires interaction of Tat with the bilayers. Figure 2.31c shows that A_L increases with increasing Tat concentration, by both model fitting and MD simulations. In a recent experimental and simulation study of the decapeptide of arginine, a similar thinning of 10 % and 12 % was observed for neutral and negatively charged bilayers, respectively [72].

It is of considerable interest to learn where Tat resides, on average, in the membrane, as this would establish a base position from which translocation would be initiated. We have combined our two main methods, MD simulations and X-ray scattering, to address this question. In general, Tats locate at the bilayer/water interface as indicated in Sect. 2.4.5, and they are close to the phosphocholine headgroup region by comparing the simulated $2z_{Tat}$ to D_{PP} in Fig. 2.31d with Fig. 2.31a. Although the SDP modeling of the X-ray data obtains excellent fits to the experimental form factors for a model with Tat deep in the hydrocarbon interior (Fig. 2.24), the corresponding simulated form factor shown in Fig. 2.30 does not fit the experimental form factor well. Figure 2.31d also shows that modeling gives smaller values for z_{Tat} than the simulation. The modeling result is supportive of the original simulation result of Herce and Garcia that Tat resides closer to the bilayer center than do the phosphocholine groups [35]. That is a base position that would be a possibly important precursor to translocation, as would the larger A_L. In a recent multi-scale simulation, it was found that arginines bind deeply to the carbonyl-glycerol groups as well as to the phosphate, while lysines bind only to the level of the phosphates [73]. This is in good agreement with our results, shown in Fig. 2.33.

Several groups have carried out calculations and MD simulations showing that the cost of moving an arginine group from water to the bilayer center is $\sim$12–26 kcal/mol [70, 74–76] or 6–7 kcal/mol if side-chain snorkeling to the surface is taken into account [77]. This is not inconsistent with our result that Tat interacts with the membrane because, as is well known, the bilayer is not just a hydrocarbon slab, but has interfacial headgroup regions where Tat can reside. It has been suggested that the free energy cost for charged amino acids entering the headgroup region is similar to that for partitioning into octanol, about an order of magnitude smaller free energy cost than partitioning into cyclohexane [78–80]. Simulations suggest that the free energy is smaller for an arginine residing in the interfacial region than in water, roughly by 3 kcal/mole, depending upon the lipid [70, 80]. Our results therefore appear energetically reasonable.

One concern with diffraction experiments on samples consisting of adjacent bilayers in a stack or in a multilamellar vesicle is that the samples have to be partially dried to obtain conventional diffraction data. But then there is no pure water layer between adjacent bilayers, so a hydrophilic peptide is forced into the interfacial, partially hydrophilic region of the lipid bilayer. In contrast, by using diffuse scattering, we obtained structure from experimental samples that had a range of lamellar D spacings (see Fig. 2.8 caption) that were considerably larger than the thickness of the bilayer in Fig. 2.31a, thereby providing an ample pure water space, typically greater than 20Å. The result that $2z_{Tat}$ shown in Fig. 2.31d is so much smaller than our repeat spacings shows that Tat preferentially associates with the membrane rather than dissociating into water.

We analyzed the secondary structures of Tats from MD simulations using the Define Secondary Structure of Proteins (DSSP) program [81]. Data from the MD simulation which has the best fit to experimental X-ray form factors show that Tat contains neither β nor α-helix structures. It appears that the membrane does not influence the conformation of solubilized Tat.

Given our structural and elastic moduli results, we now compare to other experiments in the literature. In 2008, the Wong group implicated Tat's ability to induce saddle-splay curvature with a potential role of bidentate hydrogen bonding as key [25]. Rhodamine-tagged Tat only entered GUVs when the PE headgroup was included with PS and PC lipids (PS:PC:PE, 20:40:40), indicating that hydrogen-bonding, and/or curvature-promoting lipids are required for Tat translocation. In PS:PE (20:80) lipids, they found Tat caused a highly curved cubic phase using X-ray diffraction [25]. In our experiments, there was little effect of adding DOPE to DOPC at either a 3:1 or 1:1 mole ratio on decrease in the bending modulus, bilayer thinning, or Tat's outward movement with increasing concentration. Our two results are not inconsistent, however, since curvature-promotion appears not to be required for Tat's ability to lower the energy required to bend nor to locate Tat in the bilayer, both of which may be important for Tat translocation. Yet Tat does translocate across membranes in their experiments only with PE in the membrane, so the ability to induce saddle-splay curvature may also be required for Tat's translocation. An X-ray, neutron and AFM study reported thickening upon initial Tat binding, in contradiction to our result in Fig. 2.31b that shows thinning [82]. We suggest that this difference was caused by their using stiff gel phase DPPC lipid that did not allow bound Tat to perturb the bilayer. Using a variety of techniques, including high sensitivity isothermal titration calorimetry and ^{2}H- and ^{31}P-NMR, Ziegler et al. [83] presented evidence that the lipid bilayer remains intact upon Tat binding and our results confirm this. Finally, we compare our structural results to those obtained by solid state NMR, although at a lower hydration level than in our sample. Su et al. [32] found that Tat lies parallel to the bilayer surface in the headgroup region of DMPC:DMPG (8:7) bilayers, similar to our cartoon in Fig. 2.34.

2.6 Conclusions

Although a recent MD simulation using umbrella sampling [84] found that the free energy required for R_9C to traverse a membrane was smaller if a water pore was present, we could not directly test the existence of a transient water pore from our X-ray scattering experiment. This is because, even with a water pore, the translocation process still requires crossing a free energy barrier which is a non-equilibrium process. X-ray form factors measure an equilibrium state. If the form factors obtained from water pore structures agreed well with experiments, it would indicate that the pore structure was thermodynamically stable. This may be the case for some antimicrobial peptides, but certainly not for the Tat peptide. Finding a kinetically competent pathway for the interesting phenomenon of translocation of highly charged Tat through hydrophobic membranes is difficult. An energetically passive translocation likely occurs very seldom on an MD simulation time scale, and it probably happens quickly, so it would not significantly change the average structure of the membrane in which it occurs. Although our results in this paper do not reveal a kinetically competent pathway, they do show that Tat is drawn to the surface of the membrane, and is therefore ready for translocation at a region of local thinning. And they show that these interactions tend to soften the membrane and increase the area per lipid A_L, thereby likely reducing the energy barrier for passive translocation.

Bibliography

1. R. Fischer, M. Fotin-Mleczek, H. Hufnagel, R. Brock, Break on through to the other side: biophysics and cell biology shed light on cell-penetrating peptides. ChemBioChem 6(12), 2126–2142 (2005)
2. A. Joliot, A. Prochiantz, Transduction peptides: from technology to physiology. Nat. Cell Biol. 6(3), 189–196 (2004)
3. M. Lindgren, M. Hällbrink, A. Prochiantz, U. Langel, Cell-penetrating peptides. Trends Pharmacol. Sci. 21(3), 99–103 (2000)
4. A.D. Frankel, C.O. Pabo, Cellular uptake of the Tat protein from human immunodeficiency virus. Cell 55(6), 1189–1193 (1988)
5. M. Green, P.M. Loewenstein, Autonomous functional domains of chemically synthesized human immunodeficiency virus Tat trans-activator protein. Cell 55(6), 1179–1188 (1988)
6. E. Vives, P. Brodin, B. Lebleu, HIV-1 Tat protein basic domain rapidly translocates through the plasma membrane and accumulates in the cell nucleus. J. Biolog. Chem. 272(25), 16010–16017 (1997)

7. G. Ter-Avetisyan, G. Tünnemann, D. Nowak, M. Nitschke, A. Herrmann, M. Drab, M.C. Cardoso, Cell entry of arginine-rich peptides is independent of endocytosis. J. Biolog. Chem. **284**(6), 3370–3378 (2009)
8. F. Duchardt, M. Fotin-Mleczek, H. Schwarz, R. Fischer, R. Brock, A comprehensive model for the cellular uptake of cationic cell-penetrating peptides. Traffic **8**(7), 848–866 (2007)
9. G. Tünnemann, R.M. Martin, S. Haupt, C. Patsch, F. Edenhofer, M.C. Cardoso, Cargo-dependent mode of uptake and bioavailability of Tat-containing proteins and peptides in living cells. FASEB J. **20**(11), 1775–1784 (2006)
10. A. Ziegler, P. Nervi, M. Dürrenberger, J. Seelig, The cationic cell-penetrating peptide CPPTat derived from the HIV-1 protein Tat is rapidly transported into living fibroblasts: optical, biophysical, and metabolic evidence. Biochemistry **44**(1), 138–148 (2005). PMID: 15628854.
11. J.S. Wadia, R.V. Stan, S.F. Dowdy, Transducible Tat-HA fusogenic peptide enhances escape of Tat-fusion proteins after lipid raft macropinocytosis. Nat. Med. **10**(3), 310–315 (2004)
12. I.M. Kaplan, J.S. Wadia, S.F. Dowdy, Cationic Tat peptide transduction domain enters cells by macropinocytosis. J. Control. Release **102**(1), 247–253 (2005)
13. D.A. Mann, A.D. Frankel, Endocytosis and targeting of exogenous HIV-1 Tat protein. EMBO J. **10**(7), 1733–420 (1991)
14. J.P. Richard, K. Melikov, H. Brooks, P. Prevot, B. Lebleu, L.V. Chernomordik, Cellular uptake of unconjugated Tat peptide involves clathrin-dependent endocytosis and heparan sulfate receptors. J. Biol. Chem. **280**(15), 15300–15306 (2005)
15. S.W. Jones, R. Christison, K. Bundell, C.J. Voyce, S. Brockbank, P. Newham, M.A. Lindsay, Characterisation of cell-penetrating peptide-mediated peptide delivery. Br. J. Pharmacol. **145**(8), 1093–1102 (2005)
16. A. Vendeville, F. Rayne, A. Bonhoure, N. Bettache, P. Montcourrier, B. Beaumelle, HIV-1 Tat enters T cells using coated pits before translocating from acidified endosomes and eliciting biological responses. Mol. Biol. Cell **15**(5), 2347–2360 (2004)
17. C. Foerg, U. Ziegler, J. Fernandez-Carneado, E. Giralt, R. Rennert, A.G. Beck-Sickinger, H.P. Merkle, Decoding the entry of two novel cell-penetrating peptides in HeLa cells: lipid raft-mediated endocytosis and endosomal escape. Biochemistry **44**(1), 72–81 (2005)
18. A. Fittipaldi, M. Giacca, Transcellular protein transduction using the Tat protein of HIV-1. Adv. Drug Deliv. Rev. **57**(4), 597–608 (2005)
19. Y. Liu, M. Jones, C.M. Hingtgen, G. Bu, N. Laribee, R.E. Tanzi, R.D. Moir, A. Nath, J.J. He, Uptake of HIV-1 Tat protein mediated by low-density lipoprotein receptor-related protein disrupts the neuronal metabolic balance of the receptor ligands. Nat. Med. **6**(12), 1380–1387 (2000)
20. V.P. Torchilin, R. Rammohan, V. Weissig, T.S. Levchenko, Tat peptide on the surface of liposomes affords their efficient intracellular delivery even at low temperature and in the presence of metabolic inhibitors. Proc. Natl. Acad. Sci. **98**(15), 8786–8791 (2001)
21. V.P. Torchilin, T.S. Levchenko, R. Rammohan, N. Volodina, B. Papahadjopoulos-Sternberg, G.G. D'Souza, Cell transfection in vitro and in vivo with nontoxic Tat peptide-liposome–DNA complexes. Proc. Natl. Acad. Sci. **100**(4), 1972–1977 (2003)
22. C. Rudolph, C. Plank, J. Lausier, U. Schillinger, R.H. Müller, J. Rosenecker, Oligomers of the arginine-rich motif of the HIV-1 Tat protein are capable of transferring plasmid DNA into cells. J. Biol. Chem. **278**(13), 11411–11418 (2003)
23. A. Chauhan, A. Tikoo, A.K. Kapur, M. Singh, The taming of the cell penetrating domain of the HIV Tat: myths and realities. J. Control. Release **117**(2), 148–162 (2007)
24. J. Sabatier, E. Vives, K. Mabrouk, A. Benjouad, H. Rochat, A. Duval, B. Hue, E. Bahraoui, Evidence for neurotoxic activity of Tat from human immunodeficiency virus type 1. J. Virol. **65**(2), 961–967 (1991)
25. A. Mishra, V.D. Gordon, L.H. Yang, R. Coridan, G.C.L. Wong, HIV Tat forms pores in membranes by inducing saddle-splay curvature: potential role of bidentate hydrogen bonding. Angew. Chem.-Int. Ed. **47**(16), 2986–2989 (2008)

26. S.T. Yang, E. Zaitseva, L.V. Chernomordik, K. Melikov, Cell-penetrating peptide induces leaky fusion of liposomes containing late endosome-specific anionic lipid. Biophys. J. **99**(8), 2525–2533 (2010)
27. P.E.G. Thoren, D. Persson, E.K. Esbjorner, M. Goksor, P. Lincoln, B. Norden, Membrane binding and translocation of cell-penetrating peptides. Biochemistry **43**(12), 3471–3489 (2004)
28. S. Krämer, H. Wunderli-Allenspach, No entry for Tat (44–57) into liposomes and intact mdck cells: novel approach to study membrane permeation of cell-penetrating peptides. Biochim. Biophys. Acta (BBA)-Biomembr. **1609**(2), 161–169 (2003)
29. C. Ciobanasu, J.P. Siebrasse, U. Kubitscheck, Cell-penetrating HIV-1 Tat peptides can generate pores in model membranes. Biophys. J. **99**(1), 153–62 (2010)
30. P.A. Gurnev, S.-T. Yang, K.C. Melikov, L.V. Chernomordik, S.M. Bezrukov, Cationic cell-penetrating peptide binds to planar lipid bilayers containing negatively charged lipids but does not induce conductive pores. Biophys. J. **104**(9), 1933–1939 (2013)
31. H.D. Herce, A.E. Garcia, J. Litt, R.S. Kane, P. Martin, N. Enrique, A. Rebolledo, V. Milesi, Arginine-rich peptides destabilize the plasma membrane, consistent with a pore formation translocation mechanism of cell-penetrating peptides. Biophys. J. **97**(7), 1917–1925 (2009)
32. Y.C. Su, A.J. Waring, P. Ruchala, M. Hong, Membrane-bound dynamic structure of an arginine-rich cell-penetrating peptide, the protein transduction domain of HIV Tat, from solid-state NMR. Biochemistry **49**(29), 6009–6020 (2010)
33. S. Shojania, J.D. O'Neil, HIV-1 Tat is a natively unfolded protein – the solution conformation and dynamics of reduced HIV-1 Tat-(1–72) by NMR spectroscopy. J. Biol. Chem. **281**(13), 8347–8356 (2006)
34. P. Bayer, M. Kraft, A. Ejchart, M. Westendorp, R. Frank, P. Rosch, Structural studies of HIV-1 Tat protein. J. Mol. Biol. **247**(4), 529–535 (1995)
35. H.D. Herce, A.E. Garcia, Molecular dynamics simulations suggest a mechanism for translocation of the HIV-1 Tat peptide across lipid membranes. Proc. Natl. Acad. Sci. **104**(52), 20805–20810 (2007)
36. S. Yesylevskyy, S.J. Marrink, A.E. Mark, Alternative mechanisms for the interaction of the cell-penetrating peptides penetratin and the Tat peptide with lipid bilayers. Biophys. J. **97**(1), 40–49 (2009)
37. E.D. Jarasch, C.E. Reilly, P. Comes, J. Kartenbeck, W.W. Franke, Isolation and characterization of nuclear membranes from calf and rat thymus. Hoppe Seylers Z. Physiol. Chem. **354**(8), 974–86 (1973)
38. S.A. Tristram-Nagle, Preparation of oriented, fully hydrated lipid samples for structure determination using X-ray scattering. Methods Mol. Biol. **400**, 63–75 (2007)
39. N. Kučerka, Y. Liu, N. Chu, H.I. Petrache, S. Tristram-Nagle, J.F. Nagle, Structure of fully hydrated fluid phase DMPC and DLPC lipid bilayers using X-ray scattering from oriented multilamellar arrays and from unilamellar vesicles. Biophys. J. **88**(4), 2626–2637 (2005)
40. S. Barna, M. Tate, S. Gruner, E. Eikenberry, Calibration procedures for charge-coupled device X-ray detectors. Rev. Sci. Instrum. **70**(7), 2927–2934 (1999)
41. Y. Lyatskaya, Y.F. Liu, S. Tristram-Nagle, J. Katsaras, J.F. Nagle, Method for obtaining structure and interactions from oriented lipid bilayers. Phys. Rev. E **63**(1), 0119071–0119079 (2001)
42. Y.F. Liu, J.F. Nagle, Diffuse scattering provides material parameters and electron density profiles of biomembranes. Phys. Rev. E **69**(4), 040901–040904(R) (2004)
43. Y. Liu, New method to obtain strcuture of biomembranes using diffuse X-ray scattering: application to fluid phase DOPC lipid bilayers. PhD thesis, Carnegie Mellon University, 2003
44. G. King, S. White, Determining bilayer hydrocarbon thickness from neutron diffraction measurements using strip-function models. Biophys. J. **49**(5), 1047–1054 (1986)
45. F. Heinrich, M. Lösche, Zooming in on disordered systems: Neutron reflection studies of proteins associated with fluid membranes. Biochim. Biophys. Acta (BBA) – Biomembr. **1838**(9), 2341–2349 (2014). Interfacially active peptides and proteins.

46. P. Shekhar, H. Nanda, M. Lösche, F. Heinrich, Continuous distribution model for the investigation of complex molecular architectures near interfaces with scattering techniques. J. Appl. Phys. **110**(10), 102216 (2011)

47. T. Mitsui, X-ray diffraction studies of membranes. Adv. Biophys. **10**, 97–135 (1978)

48. M.C. Wiener, R.M. Suter, J.F. Nagle, Structure of the fully hydrated gel phase of dipalmitoylphosphatidylcholine. Biophys. J. **55**(2), 315–325 (1989)

49. J.B. Klauda, N. Kučerka, B.R. Brooks, R.W. Pastor, J.F. Nagle, Simulation-based methods for interpreting x-ray data from lipid bilayers. Biophys. J. **90**(8), 2796–2807 (2006)

50. N. Kučerka, J.F. Nagle, J.N. Sachs, S.E. Feller, J. Pencer, A. Jackson, J. Katsaras, Lipid bilayer structure determined by the simultaneous analysis of neutron and X-ray scattering data. Biophys. J. **95**(5), 2356–2367 (2008)

51. S. Tristram-Nagle, Y. Liu, J. Legleiter, J.F. Nagle, Structure of gel phase DMPC determined by X-ray diffraction. Biophys. J. **83**(6), 3324–3335 (2002)

52. A.R. Braun, J.N. Sachs, J.F. Nagle, Comparing simulations of lipid bilayers to scattering data: the gromos 43A1-S3 force field. J. Phys. Chem. B **117**(17), 5065–5072 (2013)

53. http://lcgapp.cern.ch/project/cls/work-packages/mathlibs/minuit/index.html

54. B. Hess, C. Kutzner, D. van der Spoel, E. Lindahl, Gromacs 4: algorithms for highly efficient, load-balanced, scalable molecular simulation. J. Chem. Theory Comput. **4**(3), 435–447 (2008)

55. J.P.M. Jambeck, A.P. Lyubartsev, Derivation and systematic validation of a refined all-atom force field for phosphatidylcholine lipids. J. Phys. Chem. B **116**(10), 3164–3179 (2012)

56. J.P.M. Jambeck, A.P. Lyubartsev, An extension and further validation of an all-atomistic force field for biological membranes. J. Chem. Theory Comput. **8**(8), 2938–2948 (2012)

57. V. Hornak, R. Abel, A. Okur, B. Strockbine, A. Roitberg, C. Simmerling, Comparison of multiple amber force fields and development of improved protein backbone parameters. Proteins Struct. Funct. Bioinform. **65**(3), 712–725 (2006)

58. W.L. Jorgensen, J. Chandrasekhar, J.D. Madura, R.W. Impey, M.L. Klein, Comparison of simple potential functions for simulating liquid water. J. Chem. Phys. **79**(2), 926–935 (1983)

59. N. Kučerka, J. Katsaras, J. Nagle, Comparing membrane simulations to scattering experiments: Introducing the SIMtoEXP software. J. Membr. Biol. **235**(1), 43–50 (2010)

60. S. Miyamoto, P.A. Kollman, Settle: an analytical version of the shake and rattle algorithm for rigid water models. J. Comput. Chem. **13**(8), 952–962 (1992)

61. B. Hess, H. Bekker, H.J.C. Berendsen, J.G. E.M. Fraaije, Lincs: A linear constraint solver for molecular simulations. J. Comput. Chem. **18**(12), 1463–1472 (1997)

62. T. Darden, D. York, L. Pedersen, Particle mesh Ewald: an N-log(N) method for Ewald sums in large systems. J. Chem. Phys. **98**(12), 10089–10092 (1993)

63. G. Bussi, D. Donadio, M. Parrinello, Canonical sampling through velocity rescaling. J. Chem. Phys. **126**(1), 014101–420 (2007)

64. M. Parrinello, A. Rahman, Polymorphic transitions in single crystals: a new molecular dynamics method. J. Appl. Phys. **52**(12), 7182–7190 (1981)

65. N. Kučerka, S. Tristram-Nagle, J.F. Nagle, Closer look at structure of fully hydrated fluid phase DPPC bilayers. Biophysical Journal **90**(11), L83–L85 (2006)

66. N. Kučerka, S. Tristram-Nagle, J.F. Nagle, Structure of fully hydrated fluid phase lipid bilayers with monounsaturated chains. J. Membr. Biol. **208**(3), 193–202 (2005)

67. H. Petrache, S. Feller, J. Nagle, Determination of component volumes of lipid bilayers from simulations. Biophys. J. **72**(5), 2237–2242 (1997)

68. S. Tristram-Nagle, C.-P. Yang, J.F. Nagle, Thermodynamic studies of purple membrane. Biochim. Biophys. Acta (BBA)-Biomembr. **854**(1), 58–66 (1986)

69. http://www.basic.northwestern.edu/biotools/proteincalc.html

70. A.C.V. Johansson, E. Lindahl, The role of lipid composition for insertion and stabilization of amino acids in membranes. J. Chem. Phys. **130**(18), (2009)

71. S. Tristram-Nagle, J.F. Nagle, HIV-1 fusion peptide decreases bending energy and promotes curved fusion intermediates. Biophys. J. **93**(6), 2048–2055 (2007)

72. M. Vazdar, E. Wernersson, M. Khabiri, L. Cwiklik, P. Jurkiewicz, M. Hof, E. Mann, S. Kolusheva, R. Jelinek, P. Jungwirth, Aggregation of oligoarginines at phospholipid mem-

branes: molecular dynamics simulations, time-dependent fluorescence shift, and biomimetic colorimetric assays. J. Phys. Chem. B **117**(39), 11530–11540 (2013)

73. Z. Wu, Q. Cui, A. Yethiraj, Why do arginine and lysine organize lipids differently? insights from coarse-grained and atomistic simulations. J. Phys. Chem. B **117**(40), 12145–12156 (2013)

74. L.B. Li, I. Vorobyov, T.W. Allen, Potential of mean force and pk(a) profile calculation for a lipid membrane-exposed arginine side chain. J. Phys. Chem. B **112**(32), 9574–9587 (2008)

75. I. Vorobyov, L.B. Li, T.W. Allen, Assessing atomistic and coarse-grained force fields for protein-lipid interactions: the formidable challenge of an ionizable side chain in a membrane. J. Phys. Chem. B **112**(32), 9588–9602 (2008)

76. J.L. MacCallum, W.F.D. Bennett, D.P. Tieleman, Distribution of amino acids in a lipid bilayer from computer simulations. Biophys. J. **94**(9), 3393–3404 (2008)

77. E.V. Schow, J.A. Freites, P. Cheng, A. Bernsel, G. von Heijne, S.H. White, D.J. Tobias, Arginine in membranes: the connection between molecular dynamics simulations and translocon-mediated insertion experiments. J. Membr. Biol. **239**(1–2), 35–48 (2011)

78. W.C. Wimley, T.P. Creamer, S.H. White, Solvation energies of amino acid side chains and backbone in a family of host-guest pentapeptides. Biochemistry **35**(16), 5109–5124 (1996)

79. W.C. Wimley, S.H. White, Experimentally determined hydrophobicity scale for proteins at membrane interfaces. Nat. Struct. Biol. **3**(10), 842–848 (1996)

80. B. Roux, Lonely arginine seeks friendly environment. J. Gen. Physiol. **130**(2), 233–236 (2007)

81. W. Kabsch, C. Sander, Dictionary of protein secondary structure: pattern recognition of hydrogen-bonded and geometrical features. Biopolymers **22**(12), 2577–637 (1983)

82. D. Choi, J.H. Moon, H. Kim, B.J. Sung, M.W. Kim, G.Y. Tae, S.K. Satija, B. Akgun, C.J. Yu, H.W. Lee, D.R. Lee, J.M. Henderson, J.W. Kwong, K.L. Lam, K.Y.C. Lee, K. Shin, Insertion mechanism of cell-penetrating peptides into supported phospholipid membranes revealed by X-ray and neutron reflection. Soft Matter **8**(32), 8294–8297 (2012)

83. A. Ziegler, X.L. Blatter, A. Seelig, J. Seelig, Protein transduction domains of HIV-1 and SIV Tat interact with charged lipid vesicles. binding mechanism and thermodynamic analysis. Biochemistry **42**(30), 9185–94 (2003)

84. K. Huang, A.E. Garcia, Free energy of translocating an arginine-rich cell-penetrating peptide across a lipid bilayer suggests pore formation. Biophys. J. **104**(2), 412–420 (2013)

Chapter 3
Ripple Phase

Abstract This chapter presents synchrotron X-ray study of high resolution structure for the $P_{\beta'}$ ripple phase of the phospholipid dimyristoylphosphatidylcholine (DMPC). Lipid bilayers consisting of DMPC were oriented onto a silicon wafer and hydrated through the vapor in a hydration chamber. First, brief history of the ripple phase is presented. The materials and methods section describes in detail the sample preparation and experimental setups for low and wide angle X-ray scattering (LAXS and WAXS, respectively) from oriented samples. Then, I derive mathematical corrections necessary for analysis of LAXS data. The determined electron density map has a sawtooth profile similar to the result from lower resolution data, but the features are sharper allowing better estimates for the modulated bilayer profile and the distribution of headgroups along the aqueous interface. Moreover, analysis of high resolution wide angle X-ray data shows that the hydrocarbon chains in the longer, major side of the asymmetric sawtooth are packed similarly to the $L_{\beta F}$ gel phase, with chains in both monolayers coupled and tilted by $18°$ in the same direction. The absence of Bragg rods that could be associated with the minor side is consistent with disordered chains, as often suggested in the literature. I conclude with possible future experiments.

3.1 Introduction

3.1.1 Some Historical Detail

In the first structural study of the ripple phase by Tardieu et al., the crystallographic phase factor for the X-ray diffraction peaks from dilauroylphosphatidylcholine (DLPC) were obtained by a pattern recognition technique, and an electron density map was calculated [1]. In this chapter, "phase" is used to refer to two different ideas: a thermodynamic phase and a crystallographic phase factor (the crystallographic phase factor is described in Sect. 3.5). Tardieu et al. concluded that the structure corresponds to a 2D monoclinic unit cell shown in Fig. 3.1. The calculated electron density map showed that DLPC bilayers are height modulated and have asymmetric shape. The ripple wavelength λ_r was reported to be 85.3 Å, the lamellar periodicity $D = 55.3$ Å, the oblique angle $\gamma = 110°$, and ripple amplitude $A = 15$ Å.

© Springer International Publishing Switzerland 2015
K. Akabori, *Structure Determination of HIV-1 Tat/Fluid Phase Membranes and DMPC Ripple Phase Using X-Ray Scattering*, Springer Theses,
DOI 10.1007/978-3-319-22210-3_3

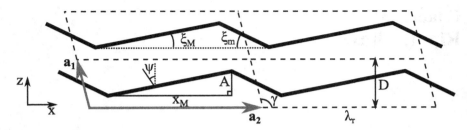

Fig. 3.1 Lattice structure of the asymmetric ripple phase. Unit cells are shown in *dashed lines*. The bilayer centers are shown by *thick, solid lines*. Notations in the figure are (a_1 and a_2: lattice unit vectors), (D: lamellar repeat distance along z), ($\lambda_r = |a_2|$: ripple wavelength), (γ: oblique angle), (A: ripple amplitude), (ψ: chain tilt angle with respect to the z-axis), and (x_M: projected length of the major arm)

Various experiments have indicated the existence of two types of ripple phases: the stable asymmetric and the metastable symmetric phase. In the asymmetric phase, there is inversion symmetry which restricts the phase factors to ± 1. In the metastable symmetric phase, there is only a plane of reflection perpendicular to the ripple wavevector a_2, so the phase factors may be complex. The metastable symmetric phase has been observed in DPPC bilayers, but not in DMPC [2, 3].

The equilibrium structure of the ripple phase in multilamellar samples has been extensively studied by X-ray diffraction [1, 4–12], neutron diffraction [13–15], freeze fracture electron microscopy [16–18], and freeze fracture scanning tunneling microscopy [19] techniques. In the scanning tunneling microscopy experiment [19], the three-dimensional contours of the ripple phase $P_{\beta'}$ of DMPC were imaged.

While many studies have used multilamellar samples, the ripple phase has been reported to also exist in large unilamellar vesicles [15, 20] and giant unilamellar vesicles (GUVs) [21]. In a GUV composed of a mixture of DPPC and dioleoylphosphatidylcholine (DOPC), coexisting domains of L'_β and P'_β have been observed [21]. However, X-ray structural studies using ULVs or GUVs are ambiguous because of the absence of out-of-plane diffraction peaks.

The ripple phase has been detected in phosphatidylcholines (PC) and phosphatidylglycerol (PG), but no ripple phase has been observed in bilayers composed entirely of phosphatidylethanolamine (PE) headgroups. These studies suggest that headgroup size influences ripple formation. Indeed, the size mismatch between the bulky PC headgroup and hydrocarbon chains lead to tilt of the chains in the gel phase [22–24]. Bilayers' tendency toward the ripple phase was also observed by including PG lipids. In the study by Li et al., where coexisting domains of L'_β and P'_β were found [21], the P'_β domain had a lower concentration of DPPC than the L'_β domain. Addition of anionic lipid DOPG caused the size of the ripple domains to grow at the expense of the gel domains. The authors concluded that reduction of surface tension drove highly stressed gel phase to less stressed ripple phase. Thus, headgroups strongly influence the formation of the ripple phase.

From X-ray data of the DMPC ripple of unoriented samples, Wack and Webb [6] argued that the ripples have a sawtooth shape but were unable to phase the observed reflections. Their intensity data were later phased by employing a modeling and fitting technique by Sun et al. [8], and the electron density map was calculated, which indicated that the ripples indeed have a sawtooth shape with a longer side called the major arm and a shorter side called the minor arm (see Fig. 3.1). The map also showed that the major arm is about twice as long as the minor arm. The major arm thickness perpendicular to the membrane plane was found to be larger than the corresponding minor arm thickness. The value of the bilayer thickness in the major arm was reported to be comparable to the thickness of DMPC bilayers in the gel phase.

Structural dependence on temperature and hydration has been studied by X-ray diffraction. The equilibrium out-of-plane structure of the DMPC ripple phase has been reported to be only weakly dependent on temperature [12]. In contrast, the ripple phase composed of POPC showed a temperature dependent structure [12]. Hydration was shown to influence the monoclinic lattice constants [6]. In multilamellar systems, the hydration level is indicated by the D-spacing, and λ_r and γ were reported to generally decrease as D increased [6]. Knowledge of λ_r, γ, A, and x_M as a function of hydration could elucidate how hydration affects the ripple sawtooth shape, but there has been no reported systematic study of the ripple structural parameters such as A and x_M as a function of hydration.

While the coarse grained electron density map of the asymmetric ripple has been well documented, the molecular organization within the bilayers has been elusive. In [11, 12], based on electron density profiles and model parameters obtained by phasing reflections from oriented samples with the modeling and fitting technique, the authors suggested that chains in both major and minor arms are tilted by the same angle with respect to the stacking z direction and that chains are nearly parallel to the local normal in the major arm. This is inconsistent with the findings in [8] that the major arm thickness is almost identical to that of the gel phase where chains are tilted by $\sim 30°$. To explain this discrepancy, they speculated that chains might be tilted by some amount into the direction perpendicular to the ripple direction.

A structural investigation by X-ray diffraction of the ripple phase of oriented dipalmitoylphosphatidylcholine (DPPC) samples indicated that hydrocarbon chains are packed in a hexagonal lattice with chains tilted in the plane perpendicular to the ripple wavevector a_2 [25]. In that study, γ was found to be $90°$. It is believed that the resolved structure was for the symmetric ripple, which has been shown to be thermodynamically metastable and whose occurrence depends on the sample history [3]. In [25], only symmetric ripple was observed in the low angle X-ray scattering, which seems contradictory to the metastability of the symmetric ripple [3].

Several works have suggested that the molecular conformation in the ripple phase consists of two distinct molecular conformations. NMR signals in the ripple phase were consistent with a superposition of signals observed in the fluid and gel phases [26]. Lateral diffusion measurements found two distinct rates, with diffusion coefficients characteristic of fluid and gel phases [27]. From these studies, the idea of micro phase separation that the major arm is gel like while the minor arm is fluid like was proposed. This idea is consistent with the later analysis [8] of the low angle X-ray scattering data from an unoriented DMPC sample [6], which revealed that the major arm is thicker than the minor arm, and the major arm thickness is comparable to the thickness of the DMPC gel phase [8]. This work was then followed by a wide angle X-ray study on unoriented samples, arguing that the micro phase separation is consistent with the wide angle data [9].

A few MD (molecular dynamics) simulations have shed light on molecular organization in the ripple phase as well. de Vries et al. [28] carried out atomistic simulations of DPPC resulting in an asymmetric ripple where chains are all-trans in the major arm but interdigitated in the minor arm. Chains in different leaflets were reported to be decoupled, the chain tilt was modulated along the ripple direction. Coarse-grain simulations performed later essentially reported the same results [29].

Many theoretical papers have been published, attempting to understand the origin of the ripple phase. A theory developed by Chen et al. [30] has been successful in describing some features in the ripple phase. In this theory, the divergence of the lipid tilt field is coupled to the curvature of the bilayer. Increase in the divergence of the lipid tilt field is compensated by an increase in the curvature, leading to the observed height modulated ripple phase. This theory predicted ripple phases with different symmetry for chiral and achiral lipids. Later, Katsaras and Raghunathan [31] carried out low angle X-ray scattering experiments on regular DMPC and achiral DMPC and found no structural difference. More recently, Kamal et al. have developed a Landau-Ginzburg theory that includes a coupling of the tilt field to the chain conformation field [32]. From their theory, fluid like chain packing was predicted in the minor arm and tilted, gel like chain packing in the major arm.

3.1.2 Purpose of This Study

Previous predictions and suggestions for molecular packing in the asymmetric ripple so far have not been directly tested because of a lack of high resolution wide angle scattering data from an oriented sample. Therefore, we sought to fill the gap with synchrotron X-ray techniques. Our strengths were three fold: (1) brilliant synchrotron beam that allowed use of Si monochromater with a very small energy dispersion, (2) stacks of ∼2000 bilayers oriented on the substrate that scattered strongly and anisotropically, and (3) hydration chamber that allowed us to control the hydration of the sample with minimum background scattering.

The symmetric ripple phase has only reflection symmetry and not centrosymmetry, so the phase factors are complex, making structure determination much more difficult than for the asymmetric ripple phase which does have centrosymmetry which restricts the phase factors to ±1. We therefore focus on the asymmetric ripple phase. We also focus on DMPC over DPPC because it is experimentally difficult to avoid coexistence of the symmetric ripple phase with the asymmetric ripple phase [3]. We chose DMPC over DLPC because the ripple phase exists in DLPC for $T < 0$ °C, which would make our experiments difficult.

The initial purpose of this study was to obtain better data relevant to the packing of the lipids within the sawtooth, asymmetric ripple profile that has been well documented [8, 12]. Chains in the major arm are believed to be packed similarly to the gel $L_{\beta'}$ phase, where chains are stretched out in the all-trans conformation. In contrast, chains in the minor arm have been suggested to be disordered like the fluid L_α phase [9, 26, 27, 33], or interdigitated like in the L_I phase [28, 29]. Structure at small length scales requires WAXS. Previously published data [31] suffer from loss of in-plane scattering intensity that we are able to obtain by using a wide angle scattering method, called transmission WAXS (tWAXS), where the x-rays go through the substrate before scattering from the sample. As it was necessary to confirm the usual low angle structural parameters for our samples, we also obtained LAXS data. Remarkably, we observed 52 well separated reflections, many more than the 17 reported reflections in the Wack and Webb data from unoriented samples [6] or the 23 reflections obtained by Sengpupta et al. from oriented samples [10]. These remarkable data are shown in Fig. 3.2, and motivated an additional project to obtain a high resolution electron density map, improving upon the previous low resolution electron density map [8].

The extraction of bilayer form factors required to obtain electron density profiles is rather more demanding for oriented samples than for unoriented samples; this is documented in Sects. 3.3 and 3.4 after describing the samples and the X-ray setup in Sect. 3.2. Obtaining the phases is also more challenging as described in Sect. 3.5 before giving final results in Sect. 3.6. The high resolution near grazing incidence WAXS (nGIWAXS) results are presented in Sect. 3.7, and the low resolution tWAXS results in Sect. 3.8. In Sect. 3.9, a model for the ripple phase WAXS pattern is developed. Structural results obtained in Sect. 3.6 are combined with the model developed in Sect. 3.9 to interpret the nGIWAXS and tWAXS data in Sect. 3.10. Section 3.10 discusses our results and compares them to previous work. We conclude this chapter in Sect. 3.12.

3.2 Materials and Methods

3.2.1 Sample Preparation

DMPC was purchased from Avanti Polar Lipids. Four mg DMPC lyophilized powder was dissolved in 140 μL chloroform:methanol (2:1 v:v) mixture. The solution was plated onto silicon wafers following the rock and roll procedure [34] (see also Sect. 2.2.3 for more details). For all the ripple phase experiments, the temperature of the hydration chamber was maintained at 18 °C. In 2011 and 2012 synchrotron experiments, the samples were created and annealed for about six hours more than a week in advance and stored in an evacuated dessicator in a refrigerator.

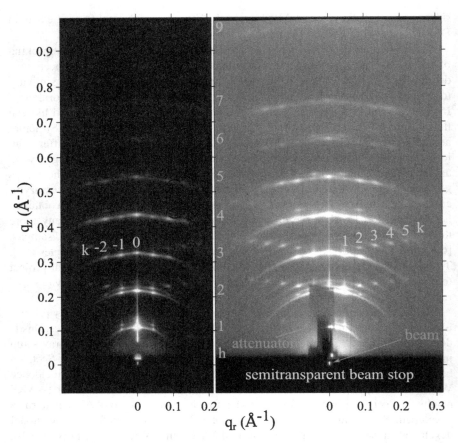

Fig. 3.2 One second exposure (*left*) and 60 s exposure (*right*) of the low angle X-ray scattering from the DMPC ripple phase in gray log intensity scales. $(3, k)$ reflections are identified. The shadow cast by 100 μm thick molybdenum attenuator blocking strong (1,0) and (2,0) orders in the *right image* is labeled as attenuator and extends from $q_z = 0$ to 0.2 Å^{-1}. The parameters defined in Fig. 3.1 have values $D = 57.8$ Å, $\lambda_r = 145.0$ Å, and $\gamma = 98.2°$

The sample quality was found to worsen over time after the samples were annealed. Therefore, to improve sample quality, in 2013 the samples were annealed for about 12 h immediately prior to the X-ray experiment. Figure 3.3 shows a picture of the annealing chamber. Annealing is promoted both by hydration and by elevated temperature. To achieve gentle but efficient hydration of a sample, filter papers were installed that exposed a larger surface for evaporation. The temperature was set to 60 °C. It must be emphasized that the annealing chamber should equilibrate in an annealing oven set to 60 °C, prior to putting a sample in the chamber. When a sample was placed in a room temperature chamber and then the system was placed inside the oven, warmer water vapor inside the chamber condensed on the cooler sample, causing so-called flooding of the oriented sample. A small drop of water on an

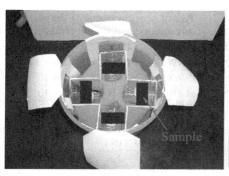

Fig. 3.3 Picture of an annealing chamber from the top (*left*) and side (*right*)

oriented film is detrimental for the orientation quality because the entropy-driven formation of ULVs causes oriented bilayers to peel off one by one and disorient.

The sample for nGIWAXS was prepared in the same way as for the low angle study. In order to minimize the geometric broadening, the sample was trimmed to 1 mm in width along the beam direction.

The sample for the tWAXS study was deposited on a thin, 35 μm thick silicon wafer and oriented following the rock and roll procedure [34]. Because the wafer was very fragile, the sample was attached to a plastic cap on a vial with a small amount of heat sink compound at a corner of the wafer. The wafer was stable enough for rocking.

3.2.2   Instrumental Resolution

The X-ray scattering experiments were carried out at the Cornell High Energy Synchrotron Source (CHESS) G1 station in three different runs (2011, 2012, and 2013). The low angle X-ray scattering (LAXS) data analyzed in this thesis were collected in 2013. The near grazing incidence wide angle X-ray scattering (nGIWAXS) data for the ripple phase were also collected in the 2013 run, but with smaller energy dispersion than in the LAXS experiment. The transmission wide angle X-ray scattering (tWAXS) data were collected in the 2011 run. The nGIWAXS data for the fluid phase were also collected in the 2011 run. The ripple phase data in the 2012 run were not used due to low sample quality. The instrumental resolution in these X-ray experiments depended on the beam divergence, energy dispersion, and geometric broadening as describe in the following subsections.

Table 3.1 Beam divergence

Year	Type of experiment	Horizontal (rad.)	Vertical (rad.)
2013	LAXS	4.2×10^{-5}	1.6×10^{-4}
2013	nGIWAXS	4.2×10^{-5}	1.6×10^{-4}
2011	tWAXS	2.5×10^{-5}	5×10^{-5}
2011	nGIWAXS	2.5×10^{-5}	5×10^{-5}

Table 3.2 Energy dispersion

Year	Type of experiment	$\Delta E/E$ (%)	E (keV)	λ (Å)
2013	LAXS	1.3	10.55	1.175
2013	nGIWAXS	0.01	10.55	1.175
2011	tWAXS	1.3	10.54	1.176
2011	nGIWAXS	1.3	10.54	1.176

3.2.2.1   Divergence

The beam divergence quantifies an angular spread of the incoming X-ray beam. We estimated the beam divergence by measuring the horizontal and vertical beam widths at two known sample-to-detector S distances with difference ΔS. The beam widths were larger at the further distance, which indicated that the beam was divergent. We calculated the divergence as $\mathrm{div} = \Delta B / \Delta S$, where ΔB is the difference in beam widths or heights at different S distances. Table 3.1 summarizes beam divergence.

3.2.2.2   Energy Dispersion

A W/B$_4$C multilayer monochromater with energy bandwidth $\Delta E/E$ of $\sim 1.3\%$ was used in the LAXS and tWAXS experiments. The energy of the X-ray beam was 10.55 keV, corresponding to an X-ray wavelength λ of 1.175 Å, in the LAXS experiment. To achieve a higher instrumental resolution, a (111) channel cut silicon monochromator was used for the nGIWAXS experiment, which gave $\Delta E/E$ of 0.01 %. Due to the geometry of the G1 station, the Si monochromator was placed in the G1 hutch, in series with the multilayer monochromator. Table 3.2 summarizes energy dispersion.

3.2.2.3   Geometric Broadening

The finite size beam footprint on the sample causes geometric broadening of diffraction peaks on the CCD detector.

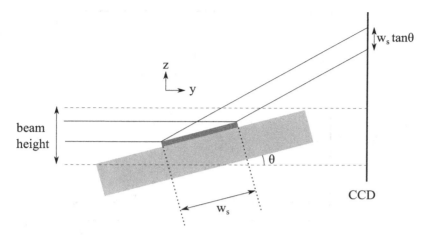

Fig. 3.4 LAXS geometry. The substrate is shown as the *gray rectangle*. The sample colored is centered in the *dashed* incoming beam. The incident angle $\omega = \theta$. The diffracted beam is broadened by $w_s \tan \theta$

LAXS

In the LAXS experiment, the geometric broadening in the horizontal x direction (see Fig. 3.4) is simply the horizontal beam width for $k = 0$ peaks with minor additional broadening for $k \neq 0$ peaks. Geometric broadening in the vertical z direction is due to different heights of the sample along the y direction of the beam at non zero angle of incidence ω. It is approximately $w_s \tan \theta$, where w_s is the sample width along the y direction and θ is the scattering angle. The beam shape, measured through a semi-transparent $200 \, \mu$m thick molybdenum (Mo) beam stop, is shown in Figs. 3.5 and 3.6. The horizontal beam width was 1.7 pixels (0.12 mm). The vertical beam height was approximately 1 mm, tall enough to cover the entire sample if the sample was tilted between $0°$ and $11.5°$. The sample was rocked during X-ray exposure between $-1.6°$ and $7°$ in order to observe many diffraction peaks in one data collection.

nGIWAXS

In near grazing incidence WAXS, the horizontal geometric broadening was due to the sample width along the beam direction and the horizontal beam width. From the geometry of the experiment shown in Fig. 3.7, the geometric broadening Δx can be estimated, assuming simple additivity,

$$\Delta x = \Delta x_{\text{beam}} + w_s \tan(2\theta),$$

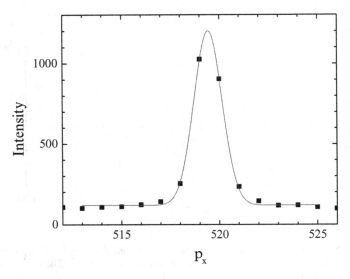

Fig. 3.5 The horizontal beam profile used in the 2013 low resolution study. The *line* is a Gaussian fit. Each pixel was 0.07113 mm, which gave a CCD angular resolution $\Delta\theta$ of 0.0057°, corresponding to $\Delta q = 0.0011\,\text{Å}^{-1}$ at the sample to detector distance of 359.7 mm. The beam FWHM = 1.7 pixels, giving $\Delta\theta = 0.010°$ or $\Delta q = 0.0019\,\text{Å}^{-1}$

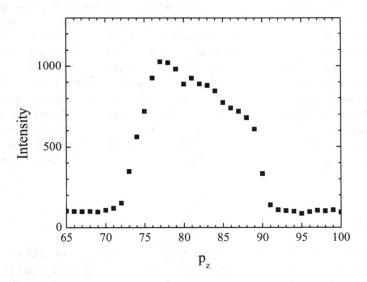

Fig. 3.6 The vertical beam profile used in the 2013 low resolution study. The beam height was 15 pixels = 1.1 mm

Fig. 3.7 In-plane geometric broadening due to the sample width w_s and the beam width Δx_{beam}. A top view of the sample on the Si wafer and the incoming and diffracted X-rays (bounded by *solid lines*) are shown. The total in-plane scattering angle is labeled 2θ, and the geometric broadening on the CCD is Δx. The sample to detector distance is not drawn to scale

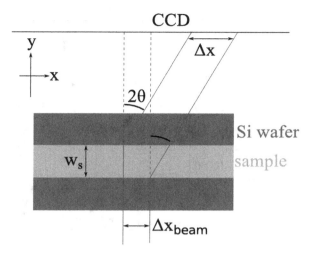

where θ is the in-plane scattering angle. The total scattering angle 2θ for the ripple WAXS was approximately $16°$. To minimize the contribution to Δx from the sample, the sample was trimmed to $w_s = 1$ mm along the beam direction. This width was chosen because (1) I could not trim more without a more sophisticated device than a simple razor blade, (2) a narrower sample would scatter less X-rays, and (3) the disordering effect from the sample edge might become too significant to ignore. The horizontal beam width was 3.7 pixels (0.26 mm) as shown in Fig. 3.8. With these experimental parameters, the resolution was estimated to be $\Delta x = 0.57$ mm $= 8$ pixels, which would be the unresolved width of an intrinsically infinitely sharp wide angle peak. Indeed, the measured width of the (2, 0) Bragg peak in the gel $L_{\beta I}$ phase was 8 pixels as will be shown in Fig. 3.48. The sample-to-detector distance was 220.6 mm, measured using silver behenate. Then, the minimum peak width measured in q-space would be $\Delta q \approx 0.014\,\text{Å}^{-1}$. The vertical geometric broadening was negligible because the sample width w_s was narrow and scattering of interest occurred at small q_z (Fig. 3.9).

tWAXS

In transmission WAXS, geometric broadening in both x and z directions was non-negligible. To calculate the broadening, let us assume that the beam has a rectangular cross section with its height Y_b and width X_b as shown in Fig. 3.10. When the sample is tilted by ω, X-rays emerging from the top edge of the sample travel a longer distance to the detector compared to the X-rays from the bottom edge of the sample. This leads to distortion of the scattered beam; namely, the scattered beam will appear on the CCD screen as a parallelogram as shown in Fig. 3.10. Figure 3.11 shows the top- and side view of the projection of the beam on the sample. From simple geometry, it can be shown that $a = Y_b/\tan\omega$, $b = aX/(2S)$, $c = aZ/(2S) + Y_b/2$, and $B = \tan^{-1}(Z/S)$. Since $H = 2c$ and $W = 2b$, H and W in Fig. 3.12 are given by

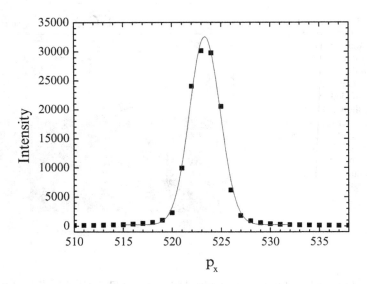

Fig. 3.8 The horizontal beam profile used in the 2013 high resolution experiment. The *line* is a Gaussian fit. The CCD angular resolution $\Delta\theta = 0.0092°$ corresponds to $\Delta q = 0.0017\,\text{Å}^{-1}$ at the sample to detector distance of 220.6 mm. The beam FWHM = 3.7 pixels = 0.26 mm, giving $\Delta\theta = 0.034°$ or $\Delta q = 0.0063\,\text{Å}^{-1}$

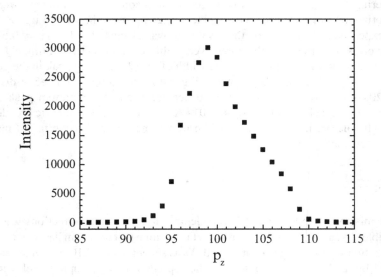

Fig. 3.9 The vertical beam profile used in the 2013 high resolution experiment. The beam height = 9 pixels = 0.64 mm

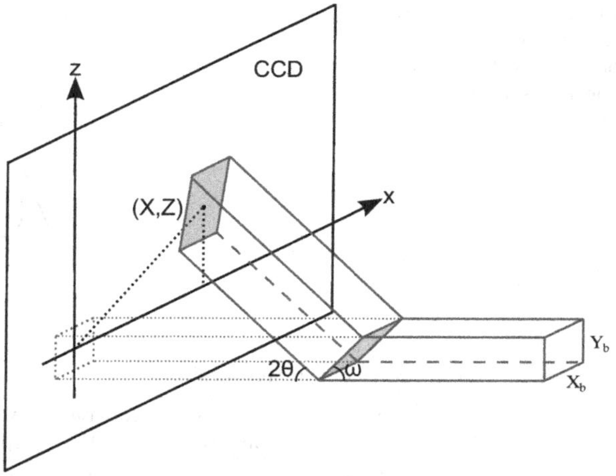

Fig. 3.10 Geometric broadening in tWAXS. The cross section of the incoming X-ray beam with the sample and the CCD detector are both shaded. The sample is tilted by $\omega = 45°$ with respect to the incoming beam. The *dots* show the transmitted beam. The incoming beam is rectangular but upon scattering appears as a parallelogram on the CCD. The sample to detector distance is not drawn to scale

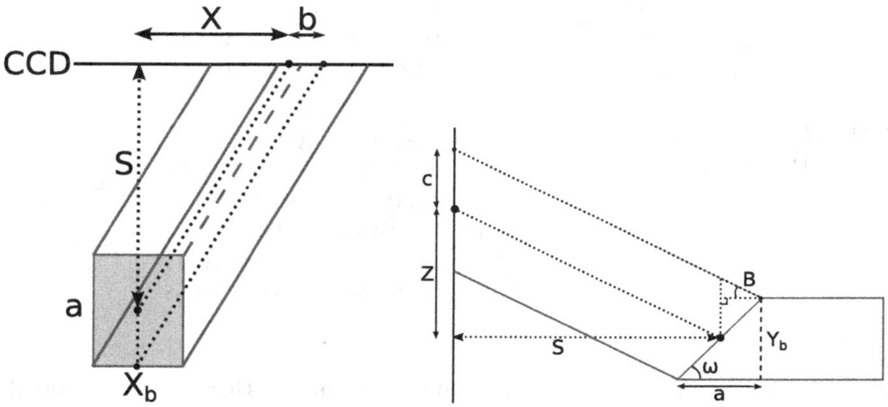

Fig. 3.11 Top (*left*) and side (*right*) views of the beam on the sample in tWAXS. The cross section of the incoming X-ray beam with the sample is shaded. X_b and Y_b are the beam width and height, respectively. S is the sample to detector distance, not drawn to scale. (X, Z) is a position of the center of the scattered beam on the detector with respect to the center of the transmitted beam as shown in Fig. 3.10

$$H = Y_b \left(1 + \frac{Z}{S \tan \omega} \right) \tag{3.1}$$

$$W = Y_b \frac{X}{S \tan \omega}. \tag{3.2}$$

The sample to detector distance S was 158.6 mm, giving an angular CCD resolution of 0.013°/pixel, or 0.0024 Å^{-1}/pixel. The observed wide angle peak was at $(X, Z) = (44.0, 15.5$ mm$)$. The beam width and height were both 0.2 mm = 2.8

Fig. 3.12 Projection of
rectangular beam on the
detector. Scattered beam
appears as a parallelogram on
the CCD

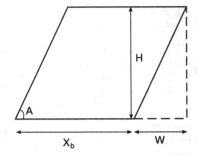

Table 3.3 Geometric broadening

Year	Type of experiment	Horizontal (pixels)	Horizontal (Å^{-1})	Vertical (pixels)	Vertical (Å^{-1})
2013	LAXS	1.7	0.0018	$6.6q_z$	$0.0070q_z$
2013	nGIWAXS	8	0.014	0	0
2011	tWAXS	2.8	0.0067	3.1	0.0074
2011	nGIWAXS	11	0.025	0	0

Table 3.4 Instrumental
resolution in q_r and q_z

Year	Type of experiment	Δq_r (Å^{-1})	Δq_z (Å^{-1})
2013	LAXS	0.0018	$0.01\sqrt{2.2q_z^2 + 0.029}$
2013	nGIWAXS	0.014	0.0017
2011	tWAXS	0.020	0.0074
2011	nGIWAXS	0.032	0.0005

pixels. With this setup, $W = 0.7$ pixels and $H = 3.1$ pixels. Therefore, the distorted
shape of the diffraction peak was negligible. Table 3.3 summarizes geometric
broadening for our experiments.

3.2.2.4 Final Instrumental Resolution

Table 3.4 lists final horizontal and vertical instrumental resolution, Δq_r and Δq_z,
combining the three contributions described in the preceding sections. The values
in Table 3.4 were calculated assuming a Gaussian resolution function for each
contribution whose full width half maximum (FWHM) is given by the estimated
resolution. and by approximating $q_r \approx 0$ for LAXS and $(q_r, q_z) \approx (1.488\,\text{Å}^{-1}, 0)$
for nGIWAXS and tWAXS.

3.2.3   Low Angle X-Ray Scattering (LAXS)

The X-ray beam for the low angle X-ray scattering (LAXS) experiment was set up by the station scientist, Dr. Arthor Woll. We chose the sample to detector distance to be 359.7 mm, measured by indexing silver behenate Bragg peaks. The D-spacing of silver behenate is known to be 58.367 Å.

Occasionally, sheets of molybdenum (Mo), each nominally 25 μm thick were used to attenuate the incoming beam. These sheets had been installed in the G1 hutch by Dr. Woll upstream of our sample chamber. The attenuation length μ of 10.55 keV X-ray in Mo is 13.74 μm [35]. For a 25 μm thick Mo attenuator, the attenuation factor is calculated to be $[\exp(-25/13.74)]^{-1} = 6.2$. The exact attenuation factor was determined by comparing X-ray images collected with and without the attenuator, shown in Fig. 3.13. The attenuation factor of the nominally 25 μm thick Mo was found to be 6.9 for the wavelength used (1.175 Å), indicating an actual thickness of 27 μm.

Sheets of Mo were also used as a semi-transparent beam stop downstream of the sample, just outside the hydration chamber, to attenuate the beam and strong orders. To avoid saturation of CCD pixels by the very intense beam of $\sim 10^{11}$ photons/mm^2/second, 200 or 225 μm were used depending on the exposure time. Also, for long exposures 100 or 200 μm were used to attenuate strong lower orders. The longest exposure times were typically 60 or 120 s (doubled for dezingering), varying somewhat for different runs.

A few Bragg peaks in the LAXS of the ripple phase were very strong, leading to saturation of CCD pixels for data collection with a long exposure time. In order to probe a wide range of q-space, three images were taken: (1) a short, one second exposure with a nominally 25 μm Mo attenuator installed upstream of the sample to reduce the intensity of the incoming X-ray beam so that the intense $(1, 0)$ reflection did not saturate the CCD, (2) one second exposure without the beam attenuator, and (3) 60 s exposure with a 100 μm Mo strip attenuating the very intense $(1, 0)$ and $(2, 0)$ peaks. The latter two exposures are shown in Fig. 3.2. Then, the integrated intensity of the $(1, 0)$ reflection was measured from the first image. This value was multiplied by 6.9 to account for the beam attenuation and then multiplied again by 60 to scale with intensities obtained at the longest exposure time. The intensities of $(2, 0)$ and $(2, -1)$ were measured from the second image, also multiplied by 60 to account for the shorter exposure time. The intensities of the rest of the observed peaks were measured from the third image.

The integrated intensity of each peak was obtained using the Nagle lab tview software developed by Dr. Yufeng Liu [36] by defining a box around a peak and summing the intensity in the pixels that fall inside the box. The background scattering was estimated by measuring the intensity in pixels near the peak but not containing any peak tail. The choice of box size was made according to the width of each peak. Because of mosaic spread in the sample, the peaks were wider for higher

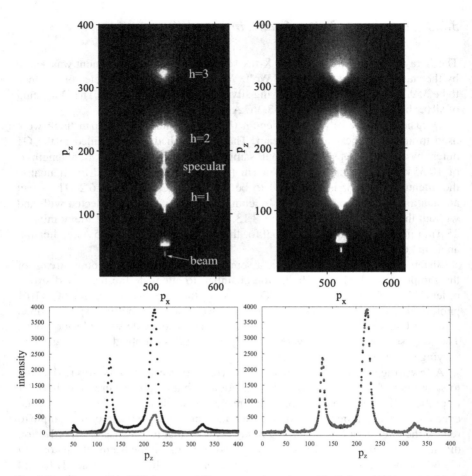

Fig. 3.13 (*Top panels*) CCD images of X-ray scattering taken with (*left*) and without (*right*) a nominally 25 μm thick Mo attenuator. These data were taken at a fixed angle of incidence $\omega = 0.8°$. The sample was an oriented film of DOPC:DOPE (3:1) in the fluid phase at 37 °C. The wavelength was 1.175 Å, the same as the one used for the ripple phase experiment. The same *gray scale* is used in both images. 100 pixel $= 0.11$ Å^{-1} in q. A *small dot* located about $(p_x, p_z) = (520, 170)$ between the first and second orders is a specular reflection from the substrate. The exposure times were 1 s. (*bottom panels*) Vertical p_z slices of the X-ray images shown in the *top panels* (*left*). The scattering intensity measured with the attenuator was multiplied by a factor of 6.9 and compared to the intensity measured without the attenuator (*right*)

orders. Consequently, the box was made wider for higher orders. The box size was chosen so that approximately 80 % of the peak intensity was counted toward the integrated intensity.

3.2.4   Near Grazing Incidence Wide Angle X-Ray Scattering (nGIWAXS)

The high resolution wide angle X-ray scattering (WAXS) experiment was also carried out at the G1 station. A channel cut silicon monochromator was set up by the G1 station scientist, Dr. Arthur Woll, and the assistant scientist, Dr. Robin Baur. WAXS was collected at an incident angle $\omega = 0.2°$. The total external reflection from an air-lipid interface occurs approximately at 0.1° and 0.17° for air-silicon interface, so 0.2° is not quite grazing incidence. Grazing incidence usually implies that the incident angle is less than the critical angle for total external reflection. Therefore, 0.2° is called near grazing incidence (nGI) in this thesis. The background scattering was collected at $\omega = -0.2°$. Subtraction of the negative angle data from the positive angle data resulted in a sample scattering image as will be shown in Fig. 3.46.

3.2.5   Transmission Wide Angle X-Ray Scattering (tWAXS)

These experiments were also carried out at the G1 station. The incident angle ω was set to $-45°$ for transmission data collection (see Fig. 3.15). A 35 μm thick silicon substrate attenuates 10.5 keV X-rays by only 20 % [35], so most of the incoming X-rays penetrated the thin substrate and none of the forward scattered X-rays were absorbed by the substrate. This is a distinct advantage of tWAXS compared to nGIWAXS because reflections with small values of q_z are not attenuated compared to those with large values of q_z.

The sample holder for tWAXS is shown in Fig. 3.14 and a schematic is shown in Fig. 3.15. Unfortunately, it was not possible to design this sample holder so that the axis of rotation of the motor in the sample chamber coincided with the sample as it does for LAXS and nGIWAXS experiments. This meant that the sample to

Fig. 3.14 Picture of the sample holder looking from above. Lead tape was attached to the back of the sample holder to help reduce the background scattering, typically coming from the air gap between the flightpath snout and the mylar window of the chamber. The sample holder was fabricated in the student machine shop

lead tape

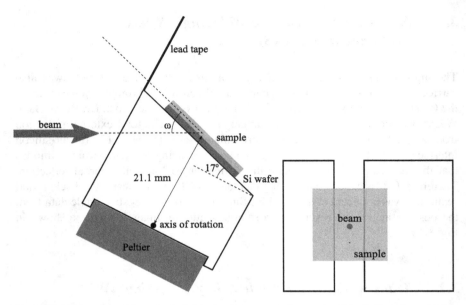

Fig. 3.15 Schematics of the sample holder in the transmission mode. Side (*left*) and top (*right*) views are shown. The Si wafer was 35 μm thick. The sample was 10 μm thick. The distance between the axis of rotation and sample = 21.1 mm

detector distance varied as ω was varied. To accurately measure the sample to detector distance, low angle scattering from a silver behenate (AgBe) sample was collected at fixed ω. Due to large mosaic spread of the AgBe sample, many orders were visible. To measure the D-spacing of the sample, ω was set to 1°. The sample to detector distance was measured to be 174.7 mm at $\omega = 0°$. From the sample holder geometry shown in Fig. 3.16, the sample to detector distance was estimated to be 158.6 mm at $\omega = 45°$.

To level the sample, the sample was first leveled coarsely by watching the sample scattering. When ω was negative, much of the incoming beam was absorbed by the flat substrate, yielding weak sample scattering. When ω became positive, sample scattering was strong. With this procedure, we leveled the sample with an uncertainty of ±0.2°. We then measured the beam intensity at various sample heights as a function of ω. The sample was level when the beam intensity had the narrowest dip as the sample was moved vertically through the beam.

Background scattering was collected by replacing the sample with a bare Si wafer. The bare Si wafer was not placed exactly at the same location as the sample, which gave slightly different background scattering. This only affected the background subtraction near the beam. The wide angle scattering was not affected by this inexact placement of the bare wafer.

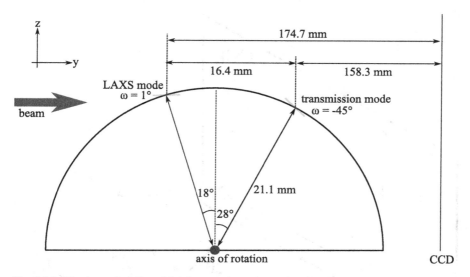

Fig. 3.16 Circular path followed by the sample as the angle of incidence ω was changed. The sample to detector distance and D-spacing of the sample were measured in the LAXS mode, where $\omega = 1°$. WAXS images were collected in the transmission mode, where $\omega = -45°$. The z position of the sample was slightly higher in the LAXS mode than in the transmission mode, so the sample holder was vertically translated for different modes. The sample to detector distance is not drawn to scale

3.2.6 Sample q-Space

The incoming and outgoing wavevectors of the X-ray beam in Fig. 3.17 are given by

$$\mathbf{k}_{in} = \frac{2\pi}{\lambda}\hat{\mathbf{y}}, \quad \mathbf{k}_{out} = \frac{2\pi}{\lambda}\left(\sin 2\theta \cos \phi\,\hat{\mathbf{x}} + \cos 2\theta\,\hat{\mathbf{y}} + \sin 2\theta \sin \phi\,\hat{\mathbf{z}}\right), \quad (3.3)$$

where λ is the X-ray wavelength, 2θ is the total scattering angle, and ϕ is the angle measured from the equator on the detector. The scattering vector (also called momentum transfer vector) is the difference between $\mathbf{k}_{out}$ and $\mathbf{k}_{in}$,

$$\mathbf{q} = \mathbf{k}_{out} - \mathbf{k}_{in}$$
$$= q\left(\cos \theta \cos \phi\,\hat{\mathbf{x}} - \sin \theta\,\hat{\mathbf{y}} + \cos \theta \sin \phi\,\hat{\mathbf{z}}\right), \quad (3.4)$$

where $q = 4\pi \sin \theta/\lambda$ is the magnitude of the scattering vector. When the sample is rotated by ω about the lab x-axis in the clockwise direction as shown in Fig. 3.17, the sample q-space also rotates and is given by

$$\hat{\mathbf{e}}_{x} = \hat{\mathbf{x}}, \quad \hat{\mathbf{e}}_{y} = \cos \omega\,\hat{\mathbf{y}} + \sin \omega\,\hat{\mathbf{z}}, \quad \hat{\mathbf{e}}_{z} = -\sin \omega\,\hat{\mathbf{y}} + \cos \omega\,\hat{\mathbf{z}}. \quad (3.5)$$

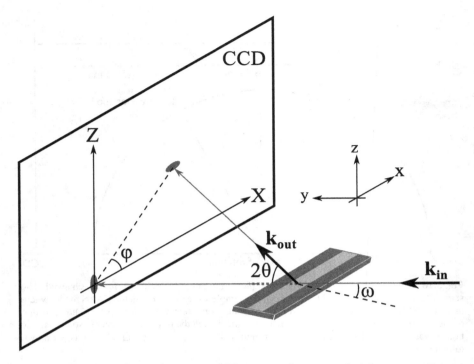

Fig. 3.17 Experimental scattering geometry. The x-, y-, and z-axes are the lab space coordinates. The incoming X-ray beam is along the y-axis with wavevector $\mathbf{k}_{in}$, and outgoing scattered X-rays make the total scattering angle 2θ with wavevector $\mathbf{k}_{out}$. The CCD detector is in the lab xz-plane. The X- and Z-axes are defined on the detector with the origin at the direct beam position. The sample is tilted by ω with respect to the incoming beam

From Eqs. (3.4) and (3.5), we find Cartesian components of the sample q-space to be

$$q_x = \mathbf{q} \cdot \hat{\mathbf{e}}_\mathbf{x} = q \cos\theta \cos\phi,$$

$$q_y = \mathbf{q} \cdot \hat{\mathbf{e}}_\mathbf{y} = q\left(-\sin\theta \cos\omega + \cos\theta \sin\phi \sin\omega\right),$$

$$q_z = \mathbf{q} \cdot \hat{\mathbf{e}}_\mathbf{z} = q\left(\sin\theta \sin\omega + \cos\theta \sin\phi \cos\omega\right). \tag{3.6}$$

The position, (X, Z), of a CCD pixel is measured with respect to the beam and given by

$$X = S \tan 2\theta \cos\phi, \quad Z = S \tan 2\theta \sin\phi, \tag{3.7}$$

where S is the distance between the sample and detector. From a model for the electron density of a lipid bilayer, one calculates the X-ray scattering intensity pattern, $I(\mathbf{q})$. Then, Eqs. (3.6) and (3.7) relate $I(\mathbf{q})$ to the experimentally measured intensity pattern, $I(X, Z)$.

For low angle x-ray scattering (LAXS), it is convenient to linearize Eq. (3.6) in terms of θ and ω. In the small angle approximation, $\sin \phi \approx Z/(2S\theta)$ and $\cos \phi \approx X/(2S\theta)$, and

$$q_x \approx \frac{4\pi\theta \cos\phi}{\lambda} \approx kX/S$$

$$q_y \approx q_z\omega - \frac{4\pi\theta^2}{\lambda} \approx q_z\omega - \frac{\lambda q_z^2}{4\pi}$$

$$q_z \approx \frac{4\pi\theta \sin\phi}{\lambda} \approx kZ/S, \tag{3.8}$$

with $k = 2\pi/\lambda$. For wide angle X-ray scattering, the exact relations given by Eq. (3.6) are necessary. Especially in the transmission experiment, where ω is large, an observed X-ray pattern appears nontrivial, and quantitative analysis requires Eq. (3.6). The transmission experiment is discussed in Sect. 3.8.

3.3 LAXS Data Reduction

The lattice structure of a stack of bilayers in the ripple phase is a two dimensional monoclinic lattice. In an oriented sample, the stacking z direction and the ripple x direction are separated, rendering peak indexing a trivial task as shown in the next subsection. However, obtaining the form factors from measured intensity is considerably more involved and requires the development of the three correction factors described in the following three subsections.

3.3.1 Lattice Structure: Unit Cell

The unit cell vectors for the two-dimensional oblique lattice shown in Fig. 3.1 can be expressed as

$$\mathbf{a_1} = \frac{D}{\tan\gamma}\hat{\mathbf{x}} + D\hat{\mathbf{z}} \tag{3.9}$$

and

$$\mathbf{a_2} = \lambda_r\hat{\mathbf{x}}. \tag{3.10}$$

The corresponding reciprocal lattice unit cell vectors are

$$\mathbf{A_1} = \frac{2\pi}{D}\hat{\mathbf{z}} \tag{3.11}$$

and

$$\mathbf{A_2} = \frac{2\pi}{\lambda_r}\hat{\mathbf{x}} - \frac{2\pi}{\lambda_r \tan \gamma}\hat{\mathbf{z}}. \tag{3.12}$$

The reciprocal lattice vector, $\mathbf{q}_{hk}$ for the Bragg peak with Miller indices (h, k) is

$$\mathbf{q}_{hk} = h\mathbf{A_1} + k\mathbf{A_2}, \tag{3.13}$$

so its Cartesian components are

$$\mathbf{q}_{hk} \cdot \hat{\mathbf{x}} = q_{hk}^x = \frac{2\pi k}{\lambda_r} \tag{3.14}$$

$$\mathbf{q}_{hk} \cdot \hat{\mathbf{y}} = q_{hk}^y = 0 \tag{3.15}$$

$$\mathbf{q}_{hk} \cdot \hat{\mathbf{z}} = q_{hk}^z = \frac{2\pi h}{D} - \frac{2\pi k}{\lambda_r \tan \gamma}. \tag{3.16}$$

Our sample consists of many ripple domains with a uniform distribution of in-plane directions of the ripple wavevector, $\mathbf{a_2}$ in Fig. 3.1. This means that, for any (h, k) reflection, there is always a domain that has an in-plane orientation such that quasi-elastic scattering occurs and a peak is observed on the CCD. In this case, q_{hk}^x and q_{hk}^y may be combined to give $q_{hk}^r = 2\pi k/\lambda_r$. Figure 3.2 shows this Miller index pattern from which the in-plane ripple repeat distance $\lambda_r = 145.0$ Å, the out-of-plane repeat distance $D = 57.8$ Å, and the oblique angle $\gamma = 98.2°$ for that sample were easily obtained. Values of q_{hk}^r and q_{hk}^z for observed reflections are included in Tables 3.8 and 3.9.

The ripple wavelength λ_r and oblique angle γ of the DMPC ripple phase depend on hydration level [6] (see Table 3.5). The best LAXS data from an unoriented sample were reported for $D = 57.9$ Å at $T = 18\,°C$. In this thesis, to compare LAXS data from an oriented sample to the data from an unoriented sample, we studied the ripple phase LAXS at the same temperature $T = 18\,°C$ and very similar hydration level $D = 57.8$ Å. The lattice constants for the data shown in Fig. 3.2 are included in Table 3.5. The bilayer structure in the ripple phase has been shown to be independent of temperature [12], so the findings in this thesis are applicable to the DMPC ripple phase at other temperature.

3.3.2   Lorentz Correction

Our sample has in-plane rotational symmetry about the z-axis. Ignoring mosaic spread to which we will come back later, this means that the sample consists of many domains with differing ripple directions, all domains being parallel to the substrate. In sample q-space, ripple $(h, k \neq 0)$ side peaks are represented as rings centered at the meridian, or q_z-axis, while $(h, k = 0)$ main peaks are still points on

Table 3.5 Lattice constants for DMPC at $T = 18.0\,°C$ reported by Wack and Webb [6] except entries with $D = 57.8, 60.1, 61.5,$ and 64.1, which are values from my best oriented sample at $T = 18.0\,°C$. The data analyzed in this thesis are $D = 57.8$. Uncertainties in our measured values were approximately $\pm 0.1\,\text{Å}$ for D, $\pm 0.5\,\text{Å}$ for λ_r, and $\pm 0.3°$ for γ

D (Å)	λ_r (Å)	γ (°)
55.0	159.4	99.0
57.0	140.8	97.6
57.3	151.6	97.8
57.4	148.4	97.6
57.5	144.1	97.8
57.5	141.9	98.0
57.8	145.0	98.2
57.9	141.7	98.4
58.0	140.1	98.2
59.8	129.6	97.3
60.1	135.2	97.7
60.6	130.1	97.0
61.5	135.1	96.7
61.5	130.8	96.5
62.4	122.0	95.9
63.9	123.1	94.9
64.1	134.8	93.2
64.9	120.3	92.3

the meridian (see Fig. 3.18). Then, for an arbitrary incident angle ω, $(h, 0)$ peaks are not observed while side peaks are observed for a range of ω as will now be explained.

In order to capture all (h, k) peaks in one X-ray exposure, the sample was continuously rotated over a range of ω, $\Delta\omega$, about the x-axis. As a result of this rotation, the $(h, 0)$ main peaks become arcs that subtend an angle $\Delta\omega$, as shown in Fig. 3.19, with its lengths equal to $\Delta\omega q^z_{h0}$. The detector records the intersections of these arcs with the Ewald sphere [37], so the intrinsic scattering intensity of the $(h, 0)$ reflections is the product of the observed intensity, I^{obs}_{h0} with the arc length, that is,

$$I_{h0} = \Delta\omega q^z_{h0} I^{obs}_{h0}. \tag{3.17}$$

This gives the usual Lorentz correction for lamellar orders.

Now, we consider relative intensity of side peaks for a given order h. As described earlier, $(h, k \neq 0)$ side peaks are represented as rings whose radius is q^r_{hk} in the sample q-space. Because only the domains with the right ripple direction can satisfy Bragg's condition at a given fixed angle ω, the intrinsic scattering intensity in this ring is reduced by a factor of $2\pi q^r_k$ compared to the $(h, 0)$ reflections. This reduction of intensity can be nicely visualized by the Ewald sphere construction shown in Fig. 3.18, which shows that the entire rings are not intersected by the Ewald sphere at a fixed angle. Then, the intrinsic scattering intensity in a ring is

$$I_{hk \neq 0} \propto 2\pi q^r_{hk} I^{obs}_{hk}. \tag{3.18}$$

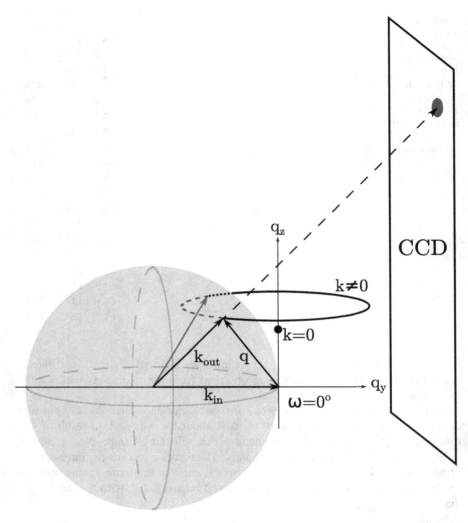

Fig. 3.18 Ewald sphere construction to obtain relation between location of scattering peaks on the CCD and their q-space values. The incoming X-ray wavevector is $\mathbf{k}_{in}$, and $\mathbf{k}_{out}$ is a scattered X-ray wavevector with $|\mathbf{k}_{out}| = |\mathbf{k}_{in}|$ for the predominant quasi-elastic scattering. A part of the q-space pattern is shown for the ripple phase in the low angle regime. For $(h, k = 0)$ Miller indices, there are points labelled $k = 0$ on the q_z axis. For $(h, k \neq 0)$ there are rings labelled $k \neq 0$ centered on the q_z-axis. The *dashed line* shows the portion of a ring that is inside the Ewald sphere and the portion outside is shown as a *black solid* or *dashed line*. Diffraction occurs where the ring and the sphere intersect. For our wavelength of 1.175 Å, $|\mathbf{k}_{in}| = 5.35\,\text{Å}^{-1}$ and for $h = 5$, $q_{50}^z = 0.54\,\text{Å}^{-1}$, one tenth of $|\mathbf{k}_{in}|$. For clarity $|\mathbf{q}|$ is drawn large compared to $|\mathbf{k}_{in}|$

Fig. 3.19 Trajectory of $k = 0$ peak as the sample is rotated by ω is shown as a *thick line*

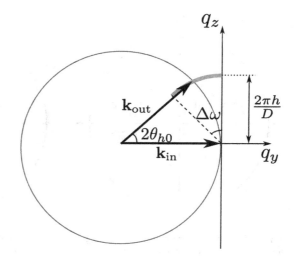

During an X-ray exposure, the sample q-space rotates and the rings are intersected by the Ewald sphere at all our experimental incident angles ω. However, as Fig. 3.20 shows, only small parts of the rings are actually intersected with the Ewald sphere. To obtain the full expression for $(h, k \neq 0)$ reflections, we now turn to a more rigorous calculation.

Mathematically, the rotation is equivalent to an integration over ω. In LAXS, q_z is nearly constant at a given pixel as ω is varied, which can be seen from Eq. (3.8). As Eq. (3.8) shows, ω dependence appears only through q_y, so rotating the sample is realized by integrating over q_y; formally, we write $d\omega = dq_y/q_z$. To derive the integration limits on q_y, let us consider two cases: (1) When $\omega \leq 0$, the incoming X-ray beam is blocked by the back of the substrate. This sets the lower limit of ω to 0. Plugging $\omega = 0$ in Eq. (3.8)), we find the lower limit of the q_y integration to be $-\lambda q_z^2/(4\pi)$. (2) When $\omega \geq 2\theta$, the substrate blocks the outgoing X-ray, so the maximum $\omega = 2\theta$. Within the small angle approximation, $q_z \approx 4\pi\theta/\lambda$. Then, the maximum ω can be expressed as $\lambda q_z/(2\pi)$. Plugging this expression for ω in Eq. (3.8), we find the upper limit of the q_y integration to be $\lambda q_z^2/(4\pi)$. Also integrating over the detector pixels X and Z to obtain integrated intensity, we write the observed intensity as

$$I_{hk}^{obs} \propto \int dX \int dZ \int d\omega\, I_{hk}$$

$$\propto \int dq_x \int dq_z \int_{-\frac{\lambda q_z^2}{4\pi}}^{\frac{\lambda q_z^2}{4\pi}} \frac{dq_y}{q_z} I_{hk}(\mathbf{q}), \qquad (3.19)$$

where $1/q_z$ factor in q_y integration is the usual Lorentz polarization factor in the small angle approximation.

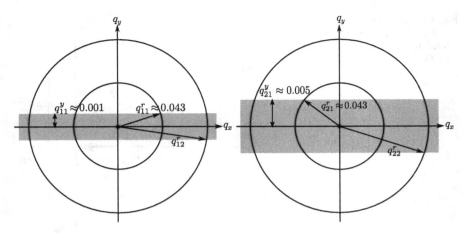

Fig. 3.20 q-space representations of Bragg peaks and Bragg rings for $h = 1$ and 2 and $k = 0, 1$, and 2 in q_{hk}^z planes. The intersection between the Ewald sphere and a Bragg peak/ring is indicated. The observed intensity for the $k \neq 0$ orders is proportional to the fraction of the length of *red arcs* to the circumference. This fraction is equal to one for $k = 0$ reflections. Because the reflections are not in the same q_z plane, the range of q_y integration indicated by the height of the *gray rectangle* is different for different h orders. For $\gamma \neq 90°$, the range of q_y integration is slightly different for different k reflections with the same h. The values shown are for $D = 58$ Å, $\lambda_r = 145$ Å, $\gamma = 90°$, and $\lambda = 1.175$ Å. For visibility, the height of the *gray rectangles* is exaggerated by about a factor of 10, exaggerating the arc curvature. With the shown large curvature, the peaks would have an asymmetric shape in the q_r direction

For a crystalline sample with in-plane rotational symmetry, the structure factor of a ripple Bragg peak is

$$S_{hk}(\mathbf{q}) = S_{hk}(q_r, q_z) = \frac{1}{2\pi q_r}\delta(q_r - q_{hk}^r)\delta(q_z - q_{hk}^z), \tag{3.20}$$

where $q_{hk}^r = 2\pi|k|/\lambda_r$. Thus, the scattering pattern in the ripple phase is a collection of Bragg rings for $k \neq 0$ centered at the meridian and the Bragg peaks for $k = 0$ located along the meridian. The scattering intensity is $I(\mathbf{q}) = |F(\mathbf{q})|^2 S(\mathbf{q})$, where $F(\mathbf{q})$ is the form factor. After the q_z integration, the observed, integrated intensity of (h, k) peak is proportional to

$$I_{hk}^{obs} \propto \frac{|F_{hk}|^2}{q_{hk}^z}\int dq_x \int_{-q_{hk}^{y0}}^{q_{hk}^{y0}} dq_y \frac{\delta(q_r - q_{hk}^r)}{2\pi q_r}, \tag{3.21}$$

where $q_{hk}^{y0} = \lambda(q_{hk}^z)^2/(4\pi)$. For side peaks ($k \neq 0$), we have

$$\int dq_x \int_{-q_{hk}^{y0}}^{q_{hk}^{y0}} dq_y \frac{\delta(q_r - q_{hk}^r)}{2\pi q_r} \approx \int_{-q_{hk}^{y0}/q_{hk}^r}^{q_{hk}^{y0}/q_{hk}^r} d\phi \int dq_r q_r \frac{\delta(q_r - q_{hk}^r)}{2\pi q_r}$$

$$= \frac{q_{hk}^{y0}}{\pi q_{hk}^r}. \tag{3.22}$$

For main peaks ($k = 0$), we have

$$\int dq_x \int_{-q_{hk}^{y0}}^{q_{hk}^{y0}} dq_y \frac{\delta(q_r - q_{hk}^r)}{2\pi q_r} = \int_0^{2\pi} d\phi \int dq_r \, q_r \frac{\delta(q_r - q_{hk}^r)}{2\pi q_r}$$

$$= 1 \tag{3.23}$$

Using Eqs. (3.21), (3.22) and (3.23), we write the observed integrated intensity as

$$I_{h0}^{\text{obs}} \propto \frac{|F_{h0}|^2}{q_{h0}^z} \tag{3.24}$$

$$I_{hk}^{\text{obs}} \propto \frac{|F_{hk}|^2}{q_{hk}^z} \frac{q_{hk}^{y0}}{\pi q_{hk}^r} = |F_{hk}|^2 \frac{\lambda q_{hk}^z}{2\pi} \frac{1}{2\pi q_{hk}^r} = |F_{hk}|^2 \frac{2\theta_{hk}}{2\pi q_{hk}^r}, \tag{3.25}$$

where $2\theta_{hk} = \lambda q_{hk}^z/(2\pi)$ is the incident angle at which the outgoing X-ray for the peak (h, k) is blocked by the substrate. Equations (3.24) and (3.25) relate the form factor calculated from a model to the experimentally observed intensity.

3.3.3 Absorption Correction for LAXS

In this section, we derive the absorption correction for an oriented sample. The calculation involves an explicit integration over the incident angle, ω, which is necessitated by the sample rotation during an X-ray exposure. The procedure is to write down an absorption factor, $A(\omega, \theta)$, for a given scattering angle 2θ at a given incident angle θ, and then integrate over ω. We ignore q_x dependence because the X-ray path inside the sample is nearly within the y-z plane for low angle scattering.

Assume that all the X-rays enter the sample from the top surface. The total scattering angle is given by 2θ (see Fig. 3.21). Let the z-axis point downward. At the

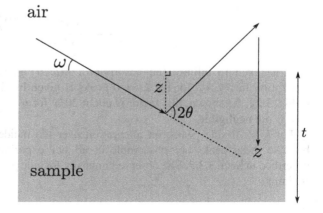

Fig. 3.21 The path of X-rays within the sample. The incident angle is ω and the total scattering angle is 2θ. An X-ray with a penetration depth of z is shown. The total thickness of the sample is t. Refraction correction is negligible for $\theta > 0.5°(h = 1)$

top surface (air-sample interface), $z = 0$. For X-rays that travel to z and then scatter, the total path length within the sample is

$$L(z, \omega, \theta) = \frac{z}{\sin \omega} + \frac{z}{\sin(2\theta - \omega)} = zg(\omega, \theta), \qquad (3.26)$$

where $g(\omega, \theta) = (\sin \omega)^{-1} + [\sin(2\theta - \omega)]^{-1}$. For each ray, the intensity is attenuated by the sample. Compared to the scattering intensity from $z = 0$, the attenuated intensity is

$$I(z, \omega, \theta) = I_0 \exp\left(-\frac{L}{\mu}\right), \qquad (3.27)$$

where μ is the absorption length of an X-ray. μ is about 2.6 mm for 10.5 keV X-ray for both water and lipids in all phases [35]. The observed sample scattering intensity at fixed ω is equal to the integration of Eq. (3.27) over the total thickness of the sample and given by

$$
\begin{aligned}
I(\omega, \theta) &= \int_0^t dz \, I(z, \omega, \theta) = I_0 \int_0^t dz \exp\left(-\frac{g(\omega, \theta)}{\mu} z\right) \\
&= I_0 \mu \frac{1 - \exp\left(-\frac{t}{\mu} g(\omega, \theta)\right)}{g(\omega, \theta)}.
\end{aligned}
\qquad (3.28)
$$

Defining the absorption factor at a fixed angle to be $A(\omega, \theta)$, the observed intensity can also be written as

$$I(\omega, \theta) = A(\omega, \theta) t I_0, \qquad (3.29)$$

where $t I_0$ is the intensity we would observe for non-absorbed X-rays. Equating Eqs. (3.28) and (3.29), we get

$$A(\omega, \theta) = \frac{\mu}{t} \frac{1 - \exp\left(-\frac{t}{\mu} g(\omega, \theta)\right)}{g(\omega, \theta)}. \qquad (3.30)$$

If μ is taken to infinity (no absorption), $A(\omega, \theta)$ goes to 1 as expected. The absorption factor A_{h0} for the $k = 0$ peaks is given by $A(\omega = \theta = \theta_B)$, plotted in Fig. 3.22. As shown, this factor is about 20 % for $h = 1$ peak relative to $h = 4$, so it is not negligible.

For $k \neq 0$ side peaks, an integration over the incident angle ω is necessary because these peaks are observable at all our experimental incident angles as described in Sect. 3.3.2. The observed intensity for side peaks from a rotating sample is simply

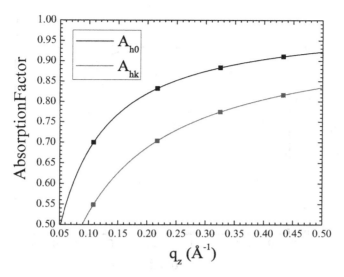

Fig. 3.22 Absorption factors Eq. (3.30) as a function of $q_z \approx 4\pi\theta/\lambda$. Values at $q_z = 2\pi h/D$ corresponding to $D = 57.8\,\text{Å}$ are shown as *squares*. $\mu = 2600\,\mu\text{m}$, $t = 10\,\mu\text{m}$, and $\lambda = 1.175\,\text{Å}$

$$I_{\text{obs}}(\theta) = \int_0^{2\theta} d\omega\, I(\omega, \theta). \tag{3.31}$$

The upper integration limit is equal to 2θ because the substrate completely blocks the scattered X-rays above this angle as discussed in Sect. 3.3.2. Equation (3.30), which is essentially the integrand in Eq. (3.31), is plotted in Fig. 3.23. It is maximum when $\omega = \theta$, meaning that the path length is shortest at the Bragg condition. The non-attenuated observed intensity is equal to $2\theta t I_0$. We define the absorption factor $A(\theta)$ to be the ratio of the total observed intensity to the total non-attenuated intensity,

$$A(\theta) \equiv \frac{I_{\text{obs}}(\theta)}{2\theta t I_0}. \tag{3.32}$$

Using Eqs. (3.30) and (3.31) in (3.32), we arrive at the final absorption factor

$$A(\theta) = \frac{1}{2\theta} \int_0^{2\theta} d\omega A(\omega, \theta) = \frac{\mu}{2\theta t} \int_0^{2\theta} d\omega \frac{1 - \exp\left(-\frac{t}{\mu} g(\omega, \theta)\right)}{g(\omega, \theta)}. \tag{3.33}$$

$A_{hk} = A(\theta)$ is plotted in Fig. 3.22. The absorption correction $A_c(\theta)$ is the inverse of Eq. (3.33).

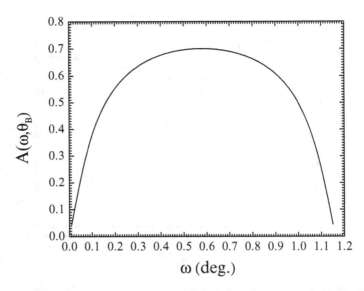

Fig. 3.23 Equation (3.30) plotted as a function of ω for $\theta = \theta_B = 0.58°$, corresponding to the $h = 1$ Bragg angle for $D = 57.8$ Å

3.3.4 Correction Due to Mosaic Spread

Integrated intensity needs to be corrected for mosaic spread, which consists of a distribution of domains of bilayers misoriented with respect to the substrate. During an X-ray exposure, the sample was continuously rotated. Due to this rotation, each pixel integrates intensity over a range of incident angles ω. As described in Appendix A.1.2, a mosaic spread distribution can be probed by changing ω, so rotating the sample is essentially equivalent to integrating a mosaic spread distribution. Because the range of the distribution probed is approximately given by $\omega = [0, 2\theta_{hk}]$ where θ_{hk} is the Bragg angle for a (h, k) reflection, this range is larger for higher h orders. This effect is illustrated in Fig. 3.24.

We limit $\chi - \chi_{hk}$ to go from $-1.4°$ to $1.4°$ by our choice of integration boxes for the intensity. The effect of the $\chi - \chi_{hk}$ cutoff is not very important because most of observed intensity was included in the integration boxes. In contrast, the cutoff on ω due to the substrate blocking the scattering is important, especially for lower h orders.

We assume the mosaic distribution to be an azimuthally symmetric 2D Lorentzian, which has been observed experimentally in this laboratory (manuscript in preparation),

$$P(\alpha) = \frac{N}{\alpha^2 + \alpha_M^2}, \tag{3.34}$$

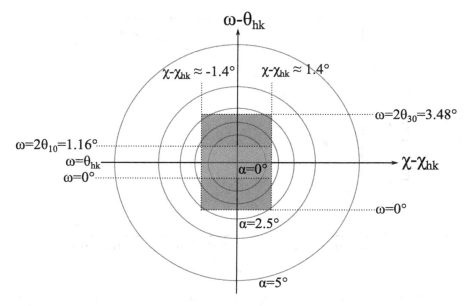

Fig. 3.24 Contours of a mosaic spread distribution projected on the $\omega\chi$-plane, where $\chi - \chi_{hk}$ is an angle measured from a (h, k) reflection on the detector ($\chi = \pi/2 - \phi$ in Fig. 3.17) and θ_{hk} is the Bragg angle for the (h, k) reflection. The distribution function is assumed to be a 2D Lorentzian centered at $\alpha = 0$. Domains with $\alpha = 0$ are probed at $\omega = \theta_{hk}$ and $\chi = \chi_{hk}$. Integrated intensity of $(1, k)$ reflections arise from domains in the *inner shaded area* while that of $(3, k)$ reflections are from the *outer shaded area*, which is three times larger

where N is a normalization constant and α_M is the half width half maximum of the distribution. N satisfies

$$N = \frac{1}{2\pi} \left(\int_0^{\frac{\pi}{2}} d\alpha \, \frac{\alpha}{\alpha^2 + \alpha_M^2} \right)^{-1}. \tag{3.35}$$

For small α, Eq. (3.34) can be approximated in terms of Cartesian coordinates as

$$P(\omega, \chi) \approx \frac{N}{\omega^2 + \chi^2 + \alpha_M^2}. \tag{3.36}$$

We then consider a two dimensional contour map on the $\omega\chi$ plane, as shown in Fig. 3.24. Mosaic factor for a reflection with Bragg angle θ_B is given by

$$M = \int_{-\theta_B}^{\theta_B} d\omega \int_{-\chi_0}^{\chi_0} d\chi \, P(\omega, \chi) = \int_{-\theta_B}^{\theta_B} d\omega \int_{-\chi_0}^{\chi_0} d\chi \, \frac{N}{\omega^2 + \chi^2 + \alpha_M^2} \tag{3.37}$$

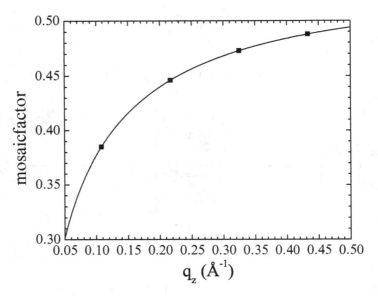

Fig. 3.25 Mosaic factor given by Eq. (3.38) as a function of $q_z \approx 4\pi\theta/\lambda$. Values at $q_z = 2\pi h/D$ corresponding to $D = 57.8$ Å are shown as *squares*. $\alpha_M = 0.05°$ and $\chi_0 = 1.4°$. Equation (3.38) reaches ~ 0.54 at $\theta_B = \pi/2$ and $\chi_0 = 1.4°$ and reaches ~ 1 at $\theta_B = \pi/2$ and $\chi_0 = \pi/2$ as expected

After the integration over χ, Eq. (3.37) is

$$M = 4N \int_0^{\theta_B} \frac{d\omega}{\sqrt{\omega^2 + \alpha_M^2}} \arctan\left(\frac{\chi_0}{\sqrt{\omega^2 + \alpha_M^2}}\right). \qquad (3.38)$$

Equation (3.38) is plotted in Fig. 3.25.

3.3.5   Synopsis of Intensity Corrections

Tables 3.6 and 3.7 show the values of the corrections obtained from the analysis in the previous three subsections using properties of our samples. The absorption and mosaicity corrections are significant for the lowest orders and their product largely accounts for the smaller intensities previously noted [38] for the lower orders of gel phase oriented samples compared to unoriented MLV samples which do not have these corrections. These two corrections decrease gradually as h increases with small modulations with k. In contrast, the Lorentz correction varies strongly with both h and k although it is the same for the same h/k. The importance of the previous three sections is emphasized by the result that the largest correction for $(1, 3)$ is a factor of 367 greater than for the smallest correction for $(1, 0)$.

Table 3.6 Correction factors for the raw intensities of the ripple LAXS peaks for thickness of an oriented sample $t = 10\,\mu m$ and mosaic spread $\alpha_M = 0.05°$

h	k	Absorption	Mosaicity	Lorentz	All
1	−1	1.96	2.63	14.16	73.086
1	0	1.41	2.56	0.11	0.394
1	1	1.79	2.56	12.67	58.027
1	2	1.74	2.53	25.00	110.055
1	3	1.69	2.50	34.12	144.592
2	−2	1.45	2.27	14.19	46.738
2	−1	1.43	2.27	6.97	22.641
2	0	1.19	2.22	0.22	0.577
2	1	1.41	2.22	6.45	20.187
2	2	1.39	2.22	12.51	38.607
2	3	1.39	2.22	18.29	56.444
2	4	1.39	2.22	23.92	73.827
2	5	1.39	2.17	28.76	86.837
2	6	1.37	2.17	33.73	100.446
3	−2	1.30	2.13	9.31	25.723
3	−1	1.30	2.13	4.50	12.436
3	0	1.14	2.13	0.33	0.788
3	1	1.28	2.08	4.35	11.586
3	2	1.28	2.08	8.52	22.766
3	3	1.28	2.08	12.56	33.555
3	4	1.27	2.08	16.42	43.295
3	5	1.27	2.08	20.18	53.212
3	6	1.27	2.08	23.81	62.802
4	−3	1.23	2.04	10.54	26.557
4	−2	1.22	2.04	6.94	17.265
4	−1	1.22	2.04	3.40	8.454
4	0	1.10	2.04	0.44	0.976
4	1	1.22	2.04	3.28	8.153
4	2	1.22	2.04	6.39	15.897
4	3	1.21	2.04	9.50	23.450
4	4	1.20	2.04	12.60	30.981
4	5	1.20	2.04	15.49	38.076
4	6	1.20	2.04	18.35	45.126

3.4 Results for $|F_{hk}|$ Form Factors

Tables 3.8 and 3.9 list the observed (h, k) reflections and their q_z and q_r values for our best sample shown in Fig. 3.2. The q_z values for observed peaks were corrected for index of refraction (Appendix A.4). Column I_{hk}^{obs} is the sum of intensity observed within an integration box centered on the peak with size shown in the box size column. These intensities were multiplied by the total correction factor redisplayed

Table 3.7 Corrections for
the intensities of the ripple
LAXS peaks (continued from
Table 3.6)

h	k	Absorption	Mosaicity	Lorentz	All
5	−3	1.19	2.00	8.44	20.084
5	−2	1.19	2.00	5.49	13.060
5	−1	1.19	2.00	2.64	6.291
5	0	1.08	2.00	0.54	1.169
5	1	1.19	2.00	2.43	5.774
6	−4	1.16	2.00	9.36	21.778
6	−3	1.16	2.00	6.92	16.094
6	−2	1.16	2.00	4.47	10.389
6	−1	1.16	2.00	2.23	5.193
6	0	1.06	2.00	0.65	1.389
6	1	1.16	2.00	2.24	5.217
6	2	1.16	2.00	4.40	10.208
6	3	1.15	2.00	6.38	14.657
6	4	1.15	2.00	8.40	19.309
7	−4	1.14	1.96	7.94	17.682
7	−3	1.14	1.96	5.86	13.060
7	−2	1.14	1.96	3.82	8.512
7	−1	1.14	1.96	1.86	4.145
7	0	1.05	1.96	0.76	1.569
8	0	1.04	1.96	0.87	1.773
9	−5	1.11	1.96	7.60	16.549
9	−4	1.11	1.96	6.07	13.233
9	−3	1.11	1.96	4.50	9.790
9	−2	1.11	1.96	2.98	6.497
9	−1	1.11	1.96	1.50	3.263
9	0	1.04	1.96	0.98	2.000

from Table 3.6, and the square root was taken to obtain unnormalized $|F_{hk}|$. As there
is an arbitrary scale factor in the data, the $|F_{hk}|$ shown in Table 3.8 were then
normalized to set $|F_{10}| = 100$.

The σ_I column in Table 3.8 gives uncertainties on I_{hk}^{obs}. The largest contribution
to σ_I for weak orders was the background scattering, which was assumed to be a
constant for each peak and estimated by plotting a swath along a given peak and
seeing where the peak tail ended. This was done visually and repeating the process
led to differences which determined the estimated σ_I. For some peaks, uncertainty
mostly came from the mosaic arc of stronger nearby peaks. For example, the (4,
−1) peak was a strong order, but the mosaic arc of its nearby stronger (4, 0) peak
overlapped with the (4, −1) peak, giving a relatively large uncertainty on the (4, −1)
peak. While most $k < 0$ peaks were susceptible to mosaic arc, $k > 0$ peaks were
not. Therefore, though $k > 0$ peaks were weaker compared to corresponding $k < 0$
peaks, their integrated intensity had smaller σ_I. We assigned a large uncertainty on
the (3, 1) peak because it overlapped with the q_z tail of the (3, −1) peak, making

Table 3.8 Observed intensity for $h = 1$–4 at $D = 57.8$, $\lambda_r = 145$, and $\gamma = 98.2°$

h	k	q_z (Å^{-1})	q_r (Å^{-1})	Box size (pixels)	I_{hk}^{obs} ($\times 10^3$)	σ_I	Correction	$|F_{hk}|$	σ_F
1	−1	0.102	−0.043	10 × 7	726.0	63.0	73.086	86.3	3.7
1	0	0.109	0.000	10 × 7	180,818.0	1759.0	0.394	100.0	0.5
1	1	0.114	0.043	10 × 7	228.0	28.0	58.027	43.1	2.6
1	2				0.0	1.0	110.055	0.0	3.9
1	3	0.128	0.130	10 × 7	3.8	0.2	144.592	8.8	0.2
2	−2	0.206	−0.087	10 × 7	49.2	3.5	46.738	18.0	0.6
2	−1	0.212	−0.044	10 × 7	1818.0	20.0	22.641	76.0	0.4
2	0	0.218	0.000	10 × 7	10,200.0	174.0	0.577	28.7	0.2
2	1	0.224	0.043	10 × 7	550.0	10.0	20.187	39.5	0.4
2	2	0.231	0.086	10 × 7	112.0	3.0	38.607	24.6	0.3
2	3	0.237	0.129	10 × 7	27.0	0.2	56.444	14.6	0.1
2	4	0.243	0.173	10 × 7	8.2	0.4	73.827	9.2	0.2
2	5	0.250	0.214	10 × 7	2.6	0.7	86.837	5.6	0.7
2	6	0.256	0.257	10 × 7	1.2	0.2	100.446	4.1	0.3
3	−2	0.314	−0.087	15 × 7	305.0	15.0	25.723	33.2	0.8
3	−1	0.321	−0.043	15 × 7	1205.0	22.0	12.436	45.9	0.4
3	0	0.326	0.000	15 × 7	1566.0	110.0	0.788	13.2	0.5
3	1			15 × 7	0.0	31.0	11.586	0.0	7.1
3	2	0.339	0.086	15 × 7	32.4	1.6	22.766	10.2	0.2
3	3	0.345	0.129	15 × 7	39.1	0.9	33.555	13.6	0.2
3	4	0.352	0.172	15 × 7	27.7	0.7	43.295	13.0	0.2
3	5	0.358	0.215	15 × 7	12.2	0.3	53.212	9.6	0.1
3	6	0.364	0.258	15 × 7	3.5	0.5	62.802	5.6	0.4
4	−3	0.417	−0.131	20 × 8	142.0	8.0	26.557	23.0	0.6
4	−2	0.423	−0.087	20 × 8	755.4	19.0	17.265	42.8	0.5
4	−1	0.429	−0.043	20 × 8	429.6	34.0	8.454	22.6	0.9
4	0	0.435	0.000	20 × 8	1917.0	23.0	0.976	16.2	0.1
4	1	0.441	0.043	20 × 8	45.3	7.2	8.153	7.2	0.6
4	2	0.448	0.085	20 × 8	43.6	2.4	15.897	9.9	0.3
4	3			20 × 8	0.0	1.3	23.450	0.0	2.1
4	4	0.461	0.173	20 × 8	2.1	0.4	30.981	3.0	0.3
4	5	0.467	0.215	20 × 8	3.2	0.3	38.076	4.1	0.2
4	6	0.473	0.259	20 × 8	1.0	1.1	45.126	2.5	1.1

separation of (3, 1) and (3, −1) difficult. It was also not clear whether the (3, 1) peak was extinct or not. σ_I for this peak was estimated by placing a box centered at the nominal position of this peak, and it is likely that a fraction of the intensity assigned to (3, −1) in Table 3.8 belongs to (3, 1). The (1, 1) and (1, −1) also overlapped in a similar manner, so their relative σ_I are larger than some of the well separated less intense peaks. Some peaks, such as (1, 2), (4, 3), (6, 2), (9, −1), (9, −3), and all the

Table 3.9 Observed intensity for $h = 5$–9 at $D = 57.8$ Å, $\lambda_r = 145$ Å, and $\gamma = 98.2°$ (continued from Table 3.8)

h	k	q_z (Å^{-1})	q_r (Å^{-1})	Box size (pixels)	I_{hk}^{obs}	σ_I	Correction	$|F_{hk}|$	σ_F
5	−3	0.525	−0.132	25 × 9	86.2	6.8	20.084	15.6	0.6
5	−2	0.532	−0.087	25 × 9	145.0	4.0	13.060	16.3	0.2
5	−1	0.538	−0.042	25 × 9	63.4	3.4	6.291	7.5	0.2
5	0	0.544	0.000	25 × 9	260.0	4.0	1.169	6.5	0.1
5	1	0.550	0.040	25 × 9	50.0	2.8	5.774	6.4	0.2
6	−4	0.628	−0.175	30 × 10	11.4	0.8	21.778	5.9	0.2
6	−3	0.635	−0.131	30 × 10	15.6	0.9	16.094	5.9	0.2
6	−2	0.641	−0.085	30 × 10	10.1	1.8	10.389	3.8	0.3
6	−1	0.647	0.043	30 × 10	16.3	3.0	5.193	3.4	0.3
6	0	0.653	0.000	30 × 10	60.2	4.7	1.389	3.4	0.1
6	1	0.659	0.044	30 × 10	20.4	1.5	5.217	3.9	0.1
6	2			30 × 10	0.0	0.6	10.208	0.0	0.9
6	3	0.672	0.128	30 × 10	5.9	0.3	14.657	3.5	0.1
6	4	0.679	0.170	30 × 10	4.2	0.3	19.309	3.4	0.1
7	−4	0.737	−0.174	35 × 10	40.0	1.1	17.682	10.0	0.1
7	−3	0.743	−0.130	35 × 10	36.0	1.8	13.060	8.1	0.2
7	−2	0.749	−0.085	35 × 10	15.0	7.3	8.512	4.2	0.9
7	−1	0.755	−0.042	35 × 10	22.0	2.3	4.145	3.6	0.2
7	0	0.760	0.000	35 × 10	36.0	1.8	1.569	2.8	0.1
8	0				0.0	3.0	1.773	0.0	0.9
9	−5	0.951	−0.215	35 × 10	16.0	3.0	16.549	6.1	0.5
9	−4	0.957	−0.173	35 × 10	16.9	3.0	13.233	5.6	0.5
9	−3			35 × 10	0.0	8.0	9.790	0.0	3.3
9	−2	0.969	−0.086	35 × 10	10.0	2.9	6.497	3.0	0.4
9	−1			35 × 10	0.0	6.0	3.263	0.0	1.7
9	0	0.981	0.000	35 × 10	17.0	10.0	2.000	2.2	0.6

$(8, k)$ peaks were deemed to be extinct because neighboring peaks had observable intensity. As zero is also an observation, these orders were also included in the table.

To assign uncertainties to the absolute form factors $|F| = \sqrt{I}$ requires propagating σ_I to σ_F. To do this, we estimated the most likely upper bound on each measured intensity $I + \sigma_I$. The most likely upper bound for $|F|$ was determined by $(|F| + \sigma_F)^2 = I + \sigma_I$, which gives σ_F,

$$\sigma_F = |F| \left(-1 + \sqrt{1 + \frac{\sigma_I}{|F|^2}} \right). \tag{3.39}$$

In the small σ_I/I regime, $\sigma_F = \sigma_I/(2|F|)$. In the large σ_I/I regime, $\sigma_F = \sqrt{\sigma_I}$. For the lower limit, a similar consideration gives the same uncertainty $\sigma_F = \sigma_I/(2|F|)$

for the small σ_I/I. The lower limit in the σ_I/I regime should be zero for the absolute form factors $|F|$. For the form factor F, we take σ_F given by Eq. (3.39) as an estimated uncertainty. For very weak peaks whose intensity could not be determined but whose nearby peaks were observed, we assigned $|F| = 0$ and $\sigma_F = \sqrt{\sigma_I}$ where σ_I was estimated based on the background scattering intensity at the q value corresponding to those unobserved weak peaks.

Our best oriented sample in Fig. 3.2 had almost the same D, γ, and only slightly different λ_r as the best data of Wack and Webb [6] from an unoriented sample. Table 3.10 compares our oriented $\left|F_{hk}^{\mathrm{ori}}\right|$ with the unoriented $\left|F_{hk}^{\mathrm{un}}\right|$. The most obvious comparison is that there are very few unoriented orders, only 18 compared to 60 orders in Tables 3.8 and 3.9. We could not determine the form factors for the $h = 0$ orders for our oriented sample because of strong attenuation of X-rays at $\omega \approx 0°$. As noted in Sun et al. [8], however, inclusion of $h = 0$ orders would not significantly alter the bilayer structure, so these orders were omitted in Table 3.8. The $k = 5$ and $k = 6$ reflections shown in Table 3.10 provide higher in-plane resolution in the oriented data, and the observability of the lamellar orders all the way to $h = 9$ provides three times higher resolution along the z-axis.

As discussed extensively in the previous section, oriented samples require complex corrections, so comparison with the relatively straightforward Lorentz correction from an unoriented sample with similar structure allows us to check our corrections. Although the ratios of the normalized form factors vary from 0.62 to 1.38, there appears to be no sign that our corrections are flawed. We propose a different reason why the ratios deviate so much from unity. In X-ray data from an oriented sample, most peaks were well separated on the two-dimensional CCD, so integrating a peak intensity was usually straightforward. In contrast, intensities from unoriented data are collapsed onto one-dimension and overlap much more, making separation of intensity difficult. Three such pairs of overlapping peaks are highlighted in Table 3.10. We show a modified $\left|F_{hk}^{\mathrm{un}}\right|$ in Table 3.10 where we have shifted some intensity from the $(1, 0)$ peak to the $(1, -1)$ peak and some intensity from the $(2, 0)$ peak to the $(2, -1)$ peak. Although there is a remaining discrepancy for the $(1, 1)$ reflection, the modified ratios are generally improved. Of course, even though it makes sense to compare these unoriented and oriented samples, one should not expect perfect agreement, especially as the ripple wavelength differs by 2.3 %.

3.5 Models to Fit the $|F_{hk}|$ and Obtain the Phase Factors

In order to obtain electron density profiles, one requires the phase factors for the $|F_{hk}|$. Once the phases are obtained, an experimental electron density map $\rho(\mathbf{r}) = \rho(x, z)$ is obtained by using

$$\rho(x, z) = \sum_{h,k} \Phi_{hk} \left|F_{hk}^{\mathrm{exp}}\right| \cos(q_{hk}^x x + q_{hk}^z z), \qquad (3.40)$$

Table 3.10 Comparison of form factors $\left|F_{hk}^{un}\right|$ for the unoriented sample from Wack and Webb [6] and $\left|F_{hk}^{ori}\right|$ from an oriented sample from this study. The ratio $\left|F_{hk}^{un}\right| / \left|F_{hk}^{ori}\right|$ of unoriented to oriented form factors is shown. Three pairs of reflections with very nearly the same q values are (0.111, 0.108), (0.215, 0.217), and (0.325, 0.325). A modification is shown that partitions the total intensity of unoriented reflections with nearly the same q, as described in the text

h	k	q (Å^{-1})	Unoriented $\left|F_{hk}^{un}\right|$	Oriented $\left|F_{hk}^{ori}\right|$	Ratio	Modified $\left|F_{hk}^{un}\right|$	Ratio
1	-1	0.111	60.8	86.3	0.70	83.0	0.96
1	0	0.108	100.0	100.0	1.00	100.0	1.00
1	1	0.123	26.9	43.1	0.62	29.9	0.69
1	2			0.0			
1	3	0.185	7.6	8.8	0.87	8.4	0.96
2	-2	0.224	15.1	18.0	0.84	16.8	0.93
2	-1	0.215	71.2	76.0	0.94	85.1	1.12
2	0	0.217	39.7	28.7	1.38	30.9	1.08
2	1	0.228	33.9	39.5	0.86	37.6	0.95
2	2	0.246	22.7	24.6	0.92	25.2	1.02
2	3	0.271	14.2	14.6	0.97	15.8	1.08
2	4	0.301	7.8	9.2	0.85	8.7	0.94
2	5	0.329		5.6			
2	6			4.1			
3	-2	0.325	29.3	33.2	0.88	32.5	0.98
3	-1	0.322	44.2	45.9	0.96	49.1	1.07
3	0	0.325	12.0	13.2	0.91	13.3	1.01
3	1			0.0			
3	2	0.350	10.5	10.2	1.03	11.7	1.15
3	3	0.370	14.9	13.6	1.10	16.5	1.22
3	4	0.394	10.0	13.0	0.77	11.1	0.86
3	5			9.6			
3	6			5.6			

where Φ_{hk} is the phase factor. Fortunately, the ripple phase has a center of inversion symmetry, so Φ_{hk} is limited to be either ±1. Nevertheless, that still leaves 2^{60} possibilities for our oriented data. It was shown by Sun et al. [8] that devising plausible models with structural parameters, such as those indicated in Fig. 3.1, and fitting those models to the observed $|F_{hk}|$ gave a robust set of phase factors for the low resolution data of Wack and Webb [6]. That strategy will be followed here.

Following Ref. [8] the electron density model for $\rho(x, z)$ within the unit cell is described as the convolution of a ripple contour function $C(x, z)$ and the transbilayer electron density profile $T_\psi(x, z)$,

$$\rho(x, z) = C(x, z) * T_\psi(x, z). \tag{3.41}$$

The form factor $F(\mathbf{q})$ is the Fourier transform of the electron density. By the convolution theorem,

$$F(\mathbf{q}) = F_C(\mathbf{q})F_T(\mathbf{q}), \tag{3.42}$$

where $F_C(\mathbf{q})$ and $F_T(\mathbf{q})$ are the Fourier transform of $C(x, z)$ and $T_\psi(x, z)$, respectively. We employed standard nonlinear least squares fitting procedures using the Levenberg-Marquardt algorithm. The software for data fitting was written in Python using the lmfit package [39].

3.5.1 Contour Part of the Form Factor

As in Ref. [8], we take the ripple profile to have a sawtooth profile. Its amplitude is A and the projection of the major arm on the ripple direction is x_M as shown in Fig. 3.1. Then, we write the ripple profile as

$$u(x) = \begin{cases} -\frac{A}{\lambda_r - x_0}\left(x + \frac{\lambda_r}{2}\right) & \text{for } -\frac{\lambda_r}{2} \le x < -\frac{x_0}{2}, \\ \frac{A}{x_0}x & \text{for } -\frac{x_0}{2} \le x \le \frac{x_0}{2}, \\ -\frac{A}{\lambda_r - x_0}\left(x - \frac{\lambda_r}{2}\right) & \text{for } \frac{x_0}{2} < x \le \frac{\lambda_r}{2}. \end{cases} \tag{3.43}$$

The ripple profile has inversion symmetry, so that the resulting form factor is real. A and x_M are fitting parameters that depend on the integrated intensity of each peak while D, λ_r, and γ are determined from measuring the positions of the Bragg peaks.

In order to allow the electron density along the ripple direction to modulate, we include two additional parameters, one to allow for the electron density across the minor side to be different by a ratio f_1 from the electron density across the major side and a second parameter f_2, which is multiplied by δ functions $\delta(x \pm x_M/2)$ to allow for a different electron density near the kink between the major and the minor sides. The full expression for the contour part of the form factor $F_C(\mathbf{q})$, which is a two dimensional Fourier transform of Eq. (3.43), is found in Appendix A.2.

3.5.2 Transbilayer Part of the Form Factor

The hybrid model developed by Wiener et al. [40] has been successful in modeling the electron density profile in the gel phase. The hybrid model with two Gaussian functions each representing the headgroup and terminal methyl group was employed by Sun et al. [8] for phasing the ripple phase X-ray data published by Wack and Webb [6]. We employed the same model for fitting our data since it was shown to be very successful in fitting the previous ripple X-ray data. Because our data contain more data points at larger q, we also used a model that has three Gaussian functions, two of which represent the headgroup and the other one represents the terminal methyl group.

In the hybrid model, the terminal methyl region of the bilayer is represented as a Gaussian function [40]. The headgroups are represented by one and two Gaussian functions in 1G and 2G hybrid models, respectively. The methylene and water regions are each treated as a constant. The gap between the two constants is represented by a sine function. Then, for half of the bilayer, $0 \leq z \leq D/2$, the electron density has the form,

$$\rho(z) = \rho_G(z) + \rho_S(z) + \rho_B(z),\tag{3.44}$$

where the Gaussian part is given by

$$\rho_G(z) = \sum_{i=1}^{1 \text{ or } 2} \rho_{Hi} e^{-(z-Z_{Hi})^2/(2\sigma_{Hi}^2)} + \rho_M e^{-z^2/(2\sigma_M^2)},\tag{3.45}$$

the strip part is given by

$$\rho_S(z) = \begin{cases} \rho_{CH_2} & \text{for} \quad 0 \leq z < Z_{CH_2}, \\ \rho_W & \text{for} \quad Z_W \leq z \leq D/2, \end{cases}\tag{3.46}$$

and the bridging part is given by

$$\rho_B(z) = \frac{\rho_W - \rho_{CH_2}}{2} \cos\left[\frac{-\pi}{\Delta Z_H}(z - Z_W)\right] + \frac{\rho_W + \rho_{CH_2}}{2} \quad \text{for} \quad Z_{CH_2} < z < Z_W\tag{3.47}$$

with $\Delta Z_H = Z_W - Z_{CH_2}$. Here, we assume $Z_{H2} > Z_{H1}$. Table 3.11 shows the definitions of Z_{CH_2} and Z_W.

The transbilayer profile along $x = -z \tan \psi$ can be obtained by rotating the coordinates x and z by ψ in the clockwise direction and reexpressing $\rho(z)$ in terms of the rotated coordinates. This leads to replacing x with $x' = x \cos \psi + z \sin \psi$ and z with $z' = -x \sin \psi + z \cos \psi$. Then, the rotated transbilayer profile is

$$\rho(x, z) = \delta(x + z \tan \psi)[\rho_G(z') + \rho_S(z') + \rho_B(z')].\tag{3.48}$$

Taking the two dimensional Fourier transform of Eq. (3.48) leads to the transbilayer part of the form factor,

Table 3.11 Definitions of Z_{CH_2} and Z_W

	1G	2G
Z_{CH_2}	$Z_{H1} - \sigma_{H1}$	$Z_{H1} - \sigma_{H1}$
Z_W	$Z_{H1} + \sigma_{H1}$	$Z_{H2} + \sigma_{H2}$

$$F_{\mathrm{T}} = \int_{-\frac{D}{2}}^{\frac{D}{2}} dz \int_{-\frac{\lambda_r}{2}}^{\frac{\lambda_r}{2}} dx [\rho(x,z) - \rho_{\mathrm{W}}] e^{i(q_x x + q_z z)} \tag{3.49}$$

$$= F_{\mathrm{G}} + F_{\mathrm{S}} + F_{\mathrm{B}}. \tag{3.50}$$

The form factor is calculated in the minus fluid convention, where the bilayer electron density is measured with respect to the electron density of the surrounding solvent, water [41]. The expression for F_{T} is rather messy, so the derivation and full expression are in Appendix A.3. Here, we note that the fitting parameters in this model are $Z_{\mathrm{H}i}$, $\sigma_{\mathrm{H}i}$, and $\rho_{\mathrm{H}i}$ for each of the two headgroup Gaussian functions, σ_{M} and ρ_{M} for the terminal methyl Gaussian, ψ for the lipid tilt, and an overall scaling factor. ρ_{CH_2} is absorbed into the overall scaling factor. The contour part of the form factor has four more parameters (A, x_{M}, f_1, and f_2). In total, the modified 2G hybrid model implements 13 structural parameters. Initially, we made $Z_{\mathrm{H}i}$, ψ, A, x_{M}, f_1, and f_2 free parameters to guide the nonlinear least squares procedure to find a reasonable fit while the other parameters were fixed to the corresponding gel phase values reported in Ref. [40]. The best estimate of the gel phase structure was reported in Ref. [42]. Precise values for the fixed parameters were not important because we then freed those parameters to find the best fit once a reasonable initial fit was obtained.

3.5.3   Some Results of Model Fitting

Table 3.12 summarizes representative fits obtained by a nonlinear least squares fitting procedure. Fit1 and Fit2 were fits using the 1G hybrid model, and Fit3–Fit7 were with the 2G hybrid model. As Table 3.12 shows, Fit5 produced the smallest χ^2 value. This fit was found by starting with Fit3, then freeing the widths of the three Gaussians (Fit4), and finally freeing the amplitudes of the Gaussians. We also tried a different route; from Fit3, we freed the amplitudes of the Gaussians (Fit6) and then freed the Gaussian widths, arriving at Fit7. We consistently obtained model form factors that were too small compared to the experimental ones for $(h, k) = (3, 0)$, $(6, k)$, and $(9, 0)$. This can be understood by inspecting the contour part of the form factor $F_C(\mathbf{q})$ given by Eq. (A.29). The model form factor $F(\mathbf{q})$ is a product of $F_C(\mathbf{q})$ and $F_T(\mathbf{q})$. Figure 3.26 plots a two dimensional map of $|F_C(\mathbf{q})|$ for $\lambda_r = 145\,\text{Å}$, $A = 21.5\,\text{Å}$, $x_{\mathrm{M}} = 103\,\text{Å}$, $f_1 = 0.5$, and $f_2 = -3$, values of which are taken from Fit5. It shows that $|F_C(\mathbf{q})|$ is very small at $(h, k) = (3, 0)$, $(6, 0)–(6,4)$, and $(9, 0)$, leading to small values of the model $F(\mathbf{q})$ for those peaks. These weak spots in $|F_C(\mathbf{q})|$ can be moved by varying A and x_{M}. However, A and x_{M} are very sensitive to strong peaks that are on the white streak in Fig. 3.26: namely, $(h, k) = (1, 0)$, $(1, -1)$, $(2, 0)$, $(2, -1)$, $(3, -1)$, $(3, -2)$, and so on. Then, for our data set, minima in the χ^2 space are normally found with values of A and x_{M} that result in $F_C(\mathbf{q})$ similar to the one shown in Fig. 3.26. This analysissuggests that better fits to those

Table 3.12 Model parameters. Fit1 and Fit2 were performed with the M1G model while Fit3–7 were with the M2G model

Model	Fit1 M1G	Fit2 M1G	Fit3 M2G	Fit4 M2G	Fit5 M2G	Fit6 M2G	Fit7 M2G
χ^2	11,996	9664	19,458	8827	8525	8905	8883
A	20.4	24.2	22.1	21.5	21.5	21.4	21.5
x_M	98.5	118.8	92.6	104.0	102.9	102.1	102.7
f_1	0.489	0.726	0.776	0.515	0.538	0.516	0.511
f_2	0[a]	−11.3	−6.06	−2.77	−2.81	−2.62	−2.63
ψ	15.2°	14.3°	10.5°	14.4°	14.4°	15.1°	14.8°
Z_{H1}	19.8	19.7	18.1	19.5	18.7	19.1	19.0
σ_{H1}	3.43[a]	3.43[a]	2.94[a]	3.06	2.51	2.94[a]	2.97
ρ_{H1}	10.77[a]	10.77[a]	9.91[a]	9.91[a]	7.03	8.38	8.45
Z_{H2}	NA	NA	20.0	20.4	22.4	23.2	23.0
σ_{H2}	NA	NA	1.47[a]	3.17	1.38	1.47[a]	1.72
ρ_{H2}	NA	NA	7.27[a]	7.27[a]	3.75	2.83	3.00
σ_M	1.67[a]	1.67[a]	1.83[a]	2.47	2.53	1.83[a]	1.87
ρ_M	9.23[a]	9.23[a]	10.9[a]	10.9[a]	5.15	6.87	6.97

[a]Parameters were fixed to the values shown

Fig. 3.26 Two dimensional map of the contour part of the form factor $|F_C(\mathbf{q})|$ given by Eq. (A.29). The color is on a log scale shown by the color bar. *Circles* are the positions of the observed peaks. The actual experimental data (Fig. 3.2) had left-right symmetry because the sample is an in-plane powder. h and k indices are labeled for some of the peaks. The experimentally observed form factors are given by the product $|F_C(\mathbf{q})||F_T(\mathbf{q})|$

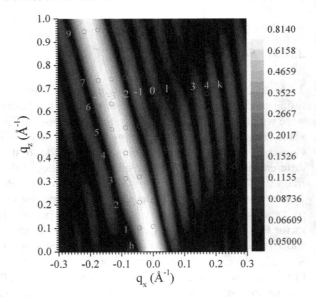

underestimated orders may require a different model for the contour part of the form factor rather than trying various models for the transbilayer part of the form factor $F_T(\mathbf{q})$. Since the sawtooth profile is a very reasonable assumption, an improvement should be made in modeling the kink regions. For example, introducing a short plateau parallel to the ripple x-axis instead of the sharp turn in the kink region of the current model would lead to a band of intensity along the q_z axis, which could bring about larger values of $|F_C(\mathbf{q})|$ at those underestimated peak positions. We did not consider improving our models because we were only interested in the predicted phases for calculating an electron density profile.

3.5.4 Results for the Phase Factors

It is important to emphasize that the goal of model fitting is to obtain the best phase factors Φ_{hk}, not to obtain the best physical values for these structural parameters. The best values of those parameters will be obtained in the next Sect. 3.6 by combining the phase factors we determine in this subsection with the experimental $|F_{hk}|$. Tables 3.13 and 3.14 show the phases that were determined by the various fits described in the previous subsection and listed in Table 3.12. The column labeled 'consensus phase' shows that the phase factor was the same for all the models for most of the reflections for which $\pm$ is entered. Reflections with an asterisk in the consensus column are extinct, so any consensus phase factor is irrelevant for the electron density profile in Eq. (3.40). We flag phase factors with a question mark as being undetermined by the models. In the case of the $(1, 3)$ reflection, there is a near consensus that $\Phi_{13} = +1$, but the model values of $|F_{13}|$ are considerably smaller than the experimental value, suggesting that Φ_{13} might have either sign. We have also flagged $\Phi_{7,-2}$ for this reason even though all models give $+1$. The most serious lack of consensus is for the $(6, k)$ reflections where the best models Fit5 and Fit7 give opposite signs. For the $(6, -3)$ reflection, both models give values of $|F_{6,-3}|$ similar in size to the experimental value which is well determined to be non-zero, but these two models give opposite $\Phi_{6,-3}$ phase factors. This emphasizes that, while the phase problem has been considerably reduced from 2^{60}, it is still necessary to consider several phase combinations to extract the best structural parameters.

3.6 Electron Density Profiles and Coarse Grained Bilayer Structure

This section concludes the LAXS analysis by presenting electron density maps and structural parameters directly obtained from the maps. Structures obtained with the form factors from the unoriented and oriented sample are compared, showing that the differences in F_{hk}^{un} and F_{hk}^{ori} in Table 3.10 are not significant for structural determination in this section. Because the phase factors for several orders were ambiguous as described in Sect. 3.5.4, our aim is to reveal robust features in the electron density maps by plotting several electron density maps using different phase factor combinations. Even with the ambiguous phase factors, the out-of-plane structure of the ripple phase was determined well as will be shown.

Table 3.13 Form factors for $h = 1\text{-}4$

h	k	Model F_{hk} Fit1	Fit2	Fit3	Fit4	Fit5	Fit6	Fit7	Consensus phase	Data $|F_{hk}|$	Error σ_F
1	−1	−74.0	−71.6	−39.4	−78.4	−77.1	−79.1	−79.8	−	86.3	3.7
1	0	−94.3	−89.2	−63.1	−98.6	−100.0	−99.6	−100.1	−	100.0	0.5
1	1	23.7	19.9	19.9	23.9	25.2	24.1	24.2	+	43.1	2.6
1	2	−6.0	−2.3	−8.3	−6.0	−6.9	−5.9	−6.0	*	0.0	3.9
1	3	0.3	−3.7	6.9	1.4	2.0	1.5	1.4	?	8.8	0.2
2	−2	−17.2	−20.2	−28.5	−19.7	−20.4	−20.1	−20.1	−	18.0	0.6
2	−1	−62.2	−59.1	−53.9	−67.9	−66.5	−65.7	−66.9	−	76.0	0.4
2	0	−32.1	−31.9	−30.8	−33.2	−33.0	−33.0	−33.1	−	28.7	0.2
2	1	31.8	30.2	32.3	31.5	31.5	32.1	32.0	+	39.5	0.4
2	2	−25.0	−24.2	−22.9	−24.0	−23.9	−24.3	−24.3	−	24.6	0.3
2	3	15.0	15.0	14.8	14.9	14.9	14.9	14.9	+	14.6	0.1
2	4	−6.1	−5.2	−12.0	−8.6	−8.9	−8.6	−8.5	−	9.2	0.2
2	5	1.1	−2.4	10.2	6.6	7.0	6.8	6.6	+	5.6	0.7
2	6	0.1	5.5	−4.0	−7.2	−7.1	−7.0	−7.0	−	4.1	0.3
3	−2	34.2	33.3	29.9	40.3	40.6	39.9	40.1	+	33.2	0.8
3	−1	39.4	39.1	27.6	45.5	44.9	44.0	44.4	+	45.9	0.4
3	0	−3.2	−4.3	−2.3	−4.3	−4.0	−4.1	−4.2	−	13.2	0.5
3	1	−9.4	−6.9	−11.2	−9.2	−9.6	−9.8	−9.5	*	0.0	7.1
3	2	14.1	12.4	15.0	14.0	14.3	14.5	14.3	+	10.2	0.2
3	3	−12.9	−13.7	−12.5	−13.1	−13.1	−13.2	−13.1	−	13.6	0.2
3	4	8.6	11.7	9.0	9.5	9.4	9.2	9.3	+	13.0	0.2
3	5	−4.1	−7.9	−7.1	−6.0	−5.9	−5.6	−5.7	−	9.6	0.1
3	6	1.1	3.6	5.4	3.9	3.9	3.6	3.7	+	5.6	0.4
4	−3	−18.1	−18.9	−18.0	−20.4	−21.7	−22.6	−21.6	−	23.0	0.6
4	−2	−48.5	−45.2	−23.9	−53.5	−53.2	−53.5	−53.0	−	42.8	0.5
4	−1	−17.8	−19.9	−7.8	−19.4	−19.0	−18.7	−18.7	−	22.6	0.9
4	0	11.3	14.3	7.8	12.7	12.6	12.7	12.6	+	16.2	0.1
4	1	−2.8	−7.8	−1.0	−4.1	−3.7	−3.7	−3.8	−	7.2	0.6
4	2	−4.0	1.6	−5.4	−2.9	−3.3	−3.5	−3.3	−	9.9	0.3
4	3	7.1	3.2	7.8	6.3	6.5	6.7	6.5	*	0.0	2.1
4	4	−6.5	−5.7	−6.8	−6.4	−6.3	−6.4	−6.4	−	3.0	0.3
4	5	4.2	6.1	5.0	4.7	4.4	4.3	4.4	+	4.1	0.2
4	6	−1.8	−4.9	−3.8	−2.8	−2.5	−2.3	−2.5	−	2.5	1.1

Figure 3.27 plots a two dimensional electron density map calculated using Eq. (3.40) with the phase factors obtained from Fit5 and our experimental form factors in Tables 3.13 and 3.14. The headgroups are electron dense and shown by white bands, which clearly indicate the sawtooth profile reported by previous X-ray diffraction studies [8, 12, 43]. Another distinct feature seen in Fig. 3.27 is the presence of the methyl trough in the major arm, manifested by a black band along

Table 3.14 Form factors for $h = 5$–9

h	k	Model F_{hk}							Consensus phase	Data $|F_{hk}|$	Error σ_F
		Fit1	Fit2	Fit3	Fit4	Fit5	Fit6	Fit7			
5	−3	−18.2	−17.8	−26.6	−16.2	−16.4	−17.7	−17.3	−	15.6	0.6
5	−2	−21.1	−21.4	−19.3	−19.3	−19.3	−19.6	−19.4	−	16.3	0.2
5	−1	1.8	1.9	4.4	2.0	2.0	2.2	2.2	+	7.5	0.2
5	0	4.7	4.8	6.4	4.3	4.6	4.5	4.3	+	6.5	0.1
5	1	−6.1	−8.3	−8.2	−6.1	−6.4	−6.3	−6.1	−	6.4	0.2
6	−4	−1.9	−1.8	6.9	2.2	2.2	−3.0	−2.8	?	5.9	0.2
6	−3	−4.3	−4.0	7.8	6.6	6.7	−5.9	−5.9	?	5.9	0.2
6	−2	−1.4	−1.7	1.5	2.7	2.8	−1.7	−1.8	?	3.8	0.3
6	−1	0.8	1.1	−2.7	−2.0	−2.2	1.1	1.1	?	3.4	0.3
6	0	−0.2	−0.5	0.8	0.7	0.7	−0.3	−0.3	?	3.4	0.1
6	1	−0.2	0.1	1.5	0.6	0.8	−0.2	−0.2	?	3.9	0.1
6	2	0.3	0.3	−2.0	−1.2	−1.5	0.3	0.3	*	0.0	0.9
6	3	−0.2	−0.5	0.5	1.0	1.2	−0.2	−0.2	?	3.5	0.1
6	4	−0.1	0.6	1.5	−0.2	−0.1	0.0	0.0	?	3.4	0.1
7	−4	−12.8	−12.0	−13.9	−9.8	−9.7	−9.6	−9.6	−	10.0	0.1
7	−3	−12.8	−13.0	−7.5	−9.6	−9.6	−9.2	−9.4	−	8.1	0.2
7	−2	1.1	0.9	3.0	0.9	1.0	1.1	1.1	?	4.2	0.9
7	−1	2.2	2.5	1.8	1.5	1.7	1.7	1.7	+	3.6	0.2
7	0	−2.4	−3.8	−3.1	−1.8	−2.1	−2.2	−2.2	−	2.8	0.1
8	0	−0.8	0.1	−1.0	−0.4	0.1	−0.4	−0.4	*	0.0	0.9
9	−5	−5.6	−5.2	2.5	−0.7	−7.3	−8.7	−8.0	−	6.1	0.5
9	−4	−5.5	−5.6	1.1	−0.6	−6.6	−8.0	−7.4	−	5.6	0.5
9	−3	0.5	0.3	−0.7	0.1	0.7	1.1	1.0	*	0.0	3.3
9	−2	0.9	1.2	−0.2	0.1	1.0	1.4	1.2	?	3.0	0.4
9	−1	−1.0	−1.7	0.7	−0.1	−1.3	−1.9	−1.7	*	0.0	1.7
9	0	0.4	1.7	−0.4	0.1	0.6	1.0	0.9	?	2.2	0.6

the bilayer center extending from $x \approx -50$ to $50\,\text{Å}$, which is not present in the minor arm. The red lines follow the electron density peak z position for the bilayer centered at $z = 0$, which define the headgroup z positions $z_{\text{Head}}(x)$ as a function of x. We define the ripple amplitude $A \equiv z_{\text{max}} - z_{\text{min}} = 18.2\,\text{Å}$ and the major arm length projected on the x-axis $x_{\text{M}} \equiv x_{\text{max}} - x_{\text{min}} = 97\,\text{Å}$ (see Fig. 3.28). These values are substantially different from the values obtained in the model fitting procedure ($A = 21.5\,\text{Å}$ and $x_{\text{M}} = 102.9\,\text{Å}$; see Table 3.12), so it is important to extract these structural parameters from the experimental electron density map. These values lead to the major arm tilt angle $\xi_{\text{M}} = 10.6°$ and the minor arm tilt angle $\xi_{\text{m}} = 20.8°$ (see Fig. 3.1 for definitions). The z_{Head} profile shows an unlikely zigzag feature in the minor arm region, suggesting that some of the phase factors obtained in Fit5 (called PF5) are incorrect. This is also seen as a sharp turn in the electron density along the headgroups $\rho_{\text{Head}}(x)$ plotted in Fig. 3.28. The ρ_{Head} profile also shows an oscillation

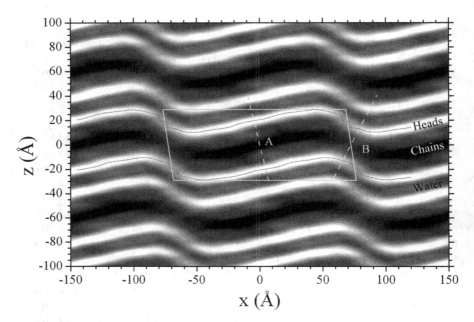

Fig. 3.27 Two dimensional electron density map calculated using Eq. (3.40) with the phase factors (PF) obtained from Fit5 (Tables 3.13 and 3.14) and our experimental form factors, in linear grayscale. *White* is most electron dense and *black* is least electron dense. A unit cell is shown with a *solid line*. *Dash lines* A and B are the slices plotted in Figs. 3.40 and 3.41, respectively. The *lines* show the locus of the highest electron density

with wavelength ~ 25 Å in the major arm region, a feature which might be removed by correcting one or more phase factors and/or measuring even more reflections.

To compare the structure obtained with the form factors from the unoriented sample $|F_{hk}^{un}|$ to the oriented sample $|F_{hk}^{ori}|$, we constructed four data sets shown in Table 3.15. The column $F_{hk}^{un,low}$ lists the form factors from the unoriented sample of Wack and Webb [6] with the phase factors reported by Sun et al. [8]; the set of (h, k) in this column will be called a low resolution set. The column $F_{hk}^{ori,high}$ is the form factors from the oriented sample with the phase factors of its low resolution set identical to the column $F_{hk}^{un,low}$ while the rest of the phase factors are identical to PF5 (see the Fit5 column in Tables 3.13 and 3.14 for the phase factors of PF5). The column $F_{hk}^{mix,high}$ was constructed by adding to the low resolution set from the unoriented sample (column $F_{hk}^{un,low}$) the form factors from the oriented sample (column $F_{hk}^{ori,high}$), and the column $F_{hk}^{ori,low}$ is the low resolution set of the column $F_{hk}^{ori,high}$.

Figure 3.29 compares z_{Head} and ρ_{Head} in the low resolution structures, obtained using the F_{hk} values in the column $F_{hk}^{un,low}$ and $F_{hk}^{ori,low}$. $A = 17$ Å and $x_M = 89$ Å were obtained using the column $F_{hk}^{un,low}$, and 18.1 Å and 91 Å using the column $F_{hk}^{ori,low}$. The low resolution structure obtained with the unoriented data set agrees

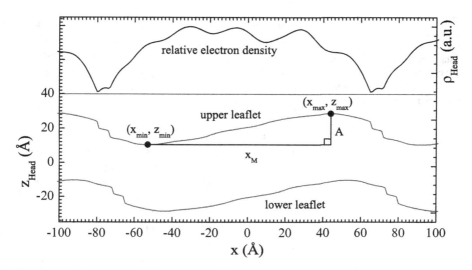

Fig. 3.28 Headgroup z positions z_{Head} as a function of x for the upper and lower leaflets, also shown as *lines* in Fig. 3.27 (PF5). The upper panel shows the electron density ρ_{Head} along the upper leaflet headgroup, also along a *line* in Fig. 3.27. *Black circles* are the points with maximum and minimum z values that define the boundaries between the major and minor arms. $A = z_{max} - z_{min}$ and $x_M = x_{max} - x_{min}$. $(x_{min}, z_{min}) = (-53, 10.4 \text{ Å})$ and $(x_{max}, z_{max}) = (44, 28.6 \text{ Å})$. $\lambda_r = 145.0 \text{ Å}$.

well with the one obtained with the oriented data set, indicating that the differences in the form factors obtained from the unoriented and oriented sample shown in Table 3.10 were not significant. The small differences in A and x_M are due to the different ripple wavelengths λ_r in the unoriented ($\lambda_r = 141.7$ Å) and oriented (145.0 Å) data sets.

Moving to the high resolution data sets, Fig. 3.30 compares the z_{Head} profile obtained using the column $F_{hk}^{mix,high}$ and $F_{hk}^{ori,high}$. It shows that the difference in z_{Head} between the unoriented and oriented data becomes negligible when the high resolution data are included. This means that the structural parameters such as A, x_M, and the major and minor arm thicknesses obtained using our measured form factors from the oriented sample are not affected by some disagreement between $|F_{hk}^{un}|$ and $|F_{hk}^{ori}|$ shown in Table 3.10. Figure 3.31 compares z_{Head} and ρ_{Head} obtained using the column $F_{hk}^{ori,low}$ and $F_{hk}^{ori,high}$. It shows that when a high resolution set is included in calculating the electron density map, major arms become longer by ~ 6 Å, leading to shorter and steeper minor arms. Table 3.16 summarizes the A and x_M values obtained from unoriented, oriented, low, and high resolution data sets listed in Table 3.15.

Figures 3.32, 3.33, 3.34 and 3.35 show electron density maps calculated using the phase factors obtained from various fits listed in Tables 3.13 and 3.14. The corresponding $z_{Head}(x)$ profiles plotted in Figs. 3.36, 3.37, 3.38 and 3.39 show a range of A and x_M, which are summarized in Table 3.17. The average values of A, x_M, ξ_M, and ξ_m are 18.5 Å, 99.2 Å, 10.5°, and 22.0°, respectively. A similar zigzag feature seen in Fig. 3.28 is present in some but not all profiles, suggesting that this

Table 3.15 Four data sets constructed to compare the unoriented to the oriented data sets. Entries for $h > 3$ in the column $F_{hk}^{\mathrm{mixed,high}}$ and $F_{hk}^{\mathrm{ori,high}}$ are identical, with the same phase factors as PF5

Sample type resolution		Unoriented low	Mixed high	Oriented low	Oriented high
h	k	$F_{hk}^{\mathrm{un,low}}$	$F_{hk}^{\mathrm{mix,high}}$	$F_{hk}^{\mathrm{ori,low}}$	$F_{hk}^{\mathrm{ori,high}}$
1	−1	−60.8	−60.8	−86.3	−86.3
1	0	−100.0	−100.0	−100.0	−100.0
1	1	26.9	26.9	43.1	43.1
1	2	*	*	*	*
1	3	7.6	7.6	8.8	8.8
2	−2	−15.1	−15.1	−18.0	−18.0
2	−1	−71.2	−71.2	−76.0	−76.0
2	0	−39.7	−39.7	−28.7	−28.7
2	1	33.9	33.9	39.5	39.5
2	2	−22.7	−22.7	−24.6	−24.6
2	3	14.2	14.2	14.6	14.6
2	4	−7.8	−7.8	−9.2	−9.2
2	5	−	5.6	−	5.6
2	6	−	−4.1	−	−4.1
3	−2	29.3	29.3	33.2	33.2
3	−1	44.2	44.2	45.9	45.9
3	0	12	12	13.2	13.2
3	1	*	*	*	*
3	2	10.5	10.5	10.2	10.2
3	3	−14.9	−14.9	−13.6	−13.6
3	4	10	10	13.0	13.0
3	5	−	−9.6	−	−9.6
3	6	−	5.6	−	5.6

Table 3.16 Summary of the parameter values obtained from oriented vs. unoriented and low vs. high resolution

	$F_{hk}^{\mathrm{un,low}}$	$F_{hk}^{\mathrm{mix,high}}$	$F_{hk}^{\mathrm{ori,low}}$	$F_{hk}^{\mathrm{ori,high}}$
A (Å)	17.0	17.7	18.1	18.1
x_M (Å)	89	98	91	97

feature may be an artifact due to incorrect phase factors. The $\rho_{\mathrm{Head}}(x)$ profiles show a plateau in the major arm except for some oscillation, and they show a dip in the minor arm.

To obtain the thickness of the bilayer in the major arm, electron density profiles calculated using the phases from various fits are plotted in Fig. 3.40 along the slice shown by the straight dashed line in Fig. 3.27 (Slice A). Slice A is along the normal of the major arm and is centered in the middle of the hydrocarbon region. It indicates that the bilayer head-head spacing $D_{\mathrm{HH}}^{\mathrm{major}}$ is 40.0–42.0 Å in the major arm (see also Table 3.17). Electron density profiles are also plotted along Slice B in Fig. 3.41. Slice B is along the normal to the minor arm and is centered in the middle of the

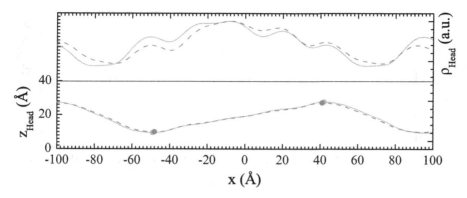

Fig. 3.29 Comparison of low resolution structures obtained using the form factors from the unoriented (*dashed*) and oriented (*solid*) samples. $z_{\text{Head}}(x)$ (*lower panel*) and $\rho_{\text{Head}}(x)$ (*upper panel*) are plotted using the F_{hk} values in the column $F_{hk}^{\text{un,low}}$ (*dashed*) and column $F_{hk}^{\text{ori,low}}$ (*solid*) in Table 3.15. The *solid circles* are the points with maximum and minimum z values that define the boundaries between the major and minor arms. $A = 17$ Å, $x_{\text{M}} = 89$ Å, and $\lambda_r = 141.7$ Å for the (*dashed*) profile, and 18.1, 91 and 145.0 Å for the (*solid*) profile

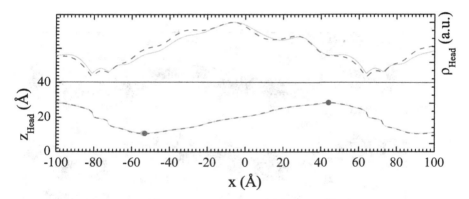

Fig. 3.30 Comparison of high resolution structures obtained using the form factors from the unoriented (*solid*) and oriented (*dashed*) samples. $z_{\text{Head}}(x)$ (*lower panel*) and $\rho_{\text{Head}}(x)$ (*upper panel*) are plotted using the F_{hk} values in the column $F_{hk}^{\text{un,high}}$ (*solid*) and column $F_{hk}^{\text{ori,high}}$ (*dashed*) in Table 3.15. The *solid circles* are the points with maximum and minimum z values that define the boundaries between the major and minor arms. $\lambda_r = 145.0$ Å was used to calculate both profiles, so that the only difference is the low resolution set F_{hk}. $A = 17.7$ Å and $x_{\text{M}} = 98$ Å for the *solid* profile and 18.1 and 97 Å for the *dashed* profile. $\lambda_r = 145.0$ Å for both profiles

hydrocarbon region. It indicates that $D_{\text{HH}}^{\text{minor}}$ is 29.2–31.0 Å in the minor arm. These results are summarized in Table 3.17.

As noted in Sect. 3.5.4, fits to $(h, k) = (3, 0)$, $(6, 0)$–$(6, 4)$, and $(9, 0)$ were unsatisfactory. We also noticed that the phase of $(1, 3)$ was unstable. To study how the electron density profile varies as the phase factors of those orders are varied, we deliberately reversed various sets of those phase factors and recalculated the

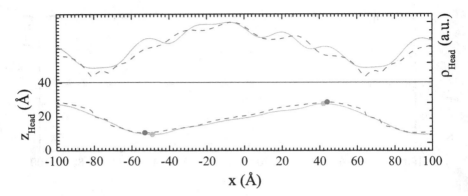

Fig. 3.31 Comparison of low (*solid*, same as in Fig. 3.29) and high (*dashed*, same as in Fig. 3.30) resolution structures. $z_{Head}(x)$ (*lower panel*) and $\rho_{Head}(x)$ (*upper panel*) are plotted using the F_{hk} values in the column $F_{hk}^{ori,low}$ (*solid*) and column $F_{hk}^{ori,high}$ (*dashed*) in Table 3.15. The *solid circles* are the points with maximum and minimum z values that define the boundaries between the major and minor arms. $A = 18.1$ Å and $x_M = 91$ Å for the *solid* profile and 18.1 and 97 Å for the *dashed* profile. $\lambda_r = 145.0$ Å for both profiles

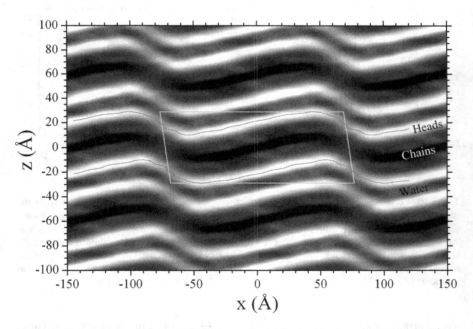

Fig. 3.32 Two dimensional electron density map calculated using the phase factors obtained from Fit1 (PF1), in linear grayscale. *White* is most electron dense and *black* is least electron dense. A unit cell is shown with a *solid line*. The *lines* show the locus of the highest electron density

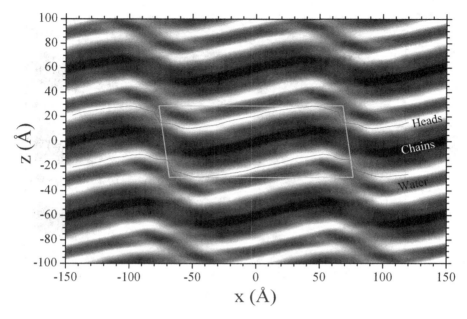

Fig. 3.33 Two dimensional electron density profile calculated using the phases factors obtained from Fit2 (PF2), in linear grayscale. *White* is most electron dense and *black* is least electron dense. A unit cell is shown with a *solid line*. The *lines* show the locus of the highest electron density

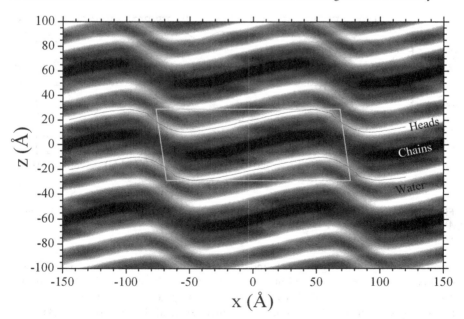

Fig. 3.34 Two dimensional electron density map calculated using the phase factors obtained from Fit3 (PF3), in linear grayscale. *White* is most electron dense and *black* is least electron dense. A unit cell is shown with a *solid line*. The *lines* show the locus of the highest electron density

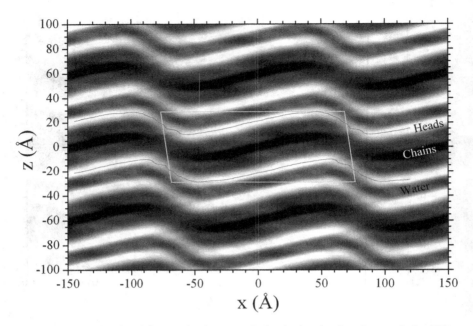

Fig. 3.35 Two dimensional electron density map calculated using the phase factors obtained from Fit7 (PF7), in linear grayscale. *White* is most electron dense and *black* is least electron dense. A unit cell is shown with a *solid line*. The *lines* show the locus of the highest electron density

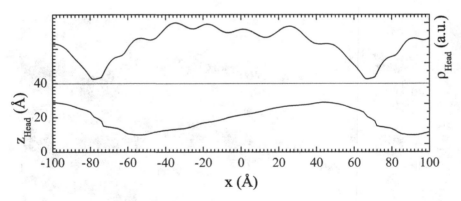

Fig. 3.36 $z_{Head}(x)$ and $\rho_{Head}(x)$ obtained from the electron density map shown in Fig. 3.32 (PF1). $A = 19.1$ Å and $x_M = 99$ Å

electron density map. Figures 3.42 and 3.43 show the major and minor arm electron density profiles for some combinations of the phases based on PF5. In PF5a, we only inverted the phase of (3, 0), and in PF5b the phases of (1, 3), (3, 0), (6, 0), and (9, 0) were inverted. In PF5c, we further reversed the sign of (6, 1–4) from PF5b. Essentially, we obtained approximately the same D_{HH}^{major} for the three cases. In

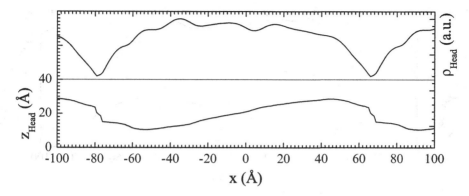

Fig. 3.37 $z_{\text{Head}}(x)$ and $\rho_{\text{Head}}(x)$ obtained from the electron density map shown in Fig. 3.33 (PF2). $A = 18.2$ Å and $x_M = 98$ Å

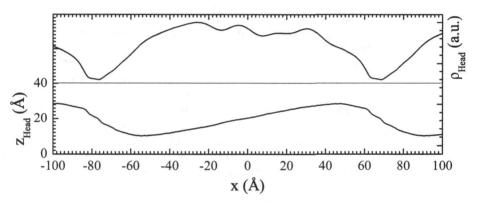

Fig. 3.38 $z_{\text{Head}}(x)$ and $\rho_{\text{Head}}(x)$ obtained from the electron density map shown in Fig. 3.34 (PF3). $A = 18.3$ Å and $x_M = 100$ Å

contrast, the variation in $D_{\text{HH}}^{\text{minor}}$ was larger, and in PF5c, $D_{\text{HH}}^{\text{minor}} = 31.8$ Å. Also, the terminal methyl trough in the major arm was present in all profiles but absent in the minor arm profiles.

In summary, we observed that the thickness of the minor arm was smaller than that of the major arm, and these thicknesses did not vary considerably considering different phase factors. Furthermore, the terminal methyl trough like feature in the major arm was quite robust, but whether the minor arm has a small dip or rise in the density at the bilayer center was not determined. Figure 3.44 plots eight major arm electron density profiles obtained with various combinations of the phase factors, and Fig. 3.45 plots equivalent curves for the minor arm. Without further analysis on the phase factors, all profiles plotted in Figs. 3.44 and 3.45 are equally possible. Elimination of one or more profiles could be done by an approach similar to pattern recognition. For example, a set of the phase factors that reproduce more physically reasonable features in the electron density map could be favored over some of the

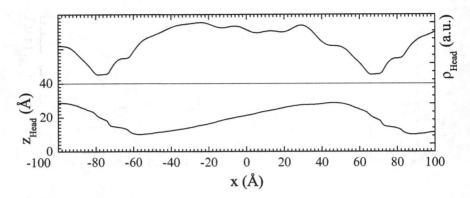

Fig. 3.39 $z_{Head}(x)$ and $\rho_{Head}(x)$ obtained from the electron density map shown in Fig. 3.35 (PF7).
$A = 18.5$ Å and $x_M = 102$ Å

Table 3.17 Structural quantities obtained using various phase factors. The units of A, x_M, D_{HH}^{major}, and D_{HH}^{minor} are Å. The $F_{hk}^{ori,low}$ column is a low resolution set from Table 3.15

	PF1	PF2	PF3	PF5	PF7	$F_{hk}^{ori,low}$
A	19.1	18.2	18.3	18.2	18.5	18.1
x_M	99	98	100	97	102	91
ξ_M	10.9°	10.5°	10.4°	10.6°	10.3°	11.2°
ξ_m	22.5°	21.2°	22.1°	20.8°	23.3°	18.5°
D_{HH}^{major}	42.0	41.2	40.3	40.7	41.8	38
D_{HH}^{minor}	30.8	31.0	29.2	29.2	31.0	31
χ^2	11,996	9664	19,458	8525	8883	

phase factors we obtained from the fits. Table 3.18 summarizes the final structural results, averaging the structural parameters measured from electron density maps presented in this section. We note that the quoted average values are not strictly statistical averages but should be reasonable estimates for the parameters.

3.7   nGIWAXS: Results

3.7.1   Fluid and Gel Phase

Figure 3.46 shows the data reduction of near grazing incidence wide angle X-ray scattering (nGIWAXS) data of the DMPC fluid phase at $T = 30\,°C$. The scattering image taken at $\omega = 0.5°$ had unwanted scattering due to mylar windows in the hydration chamber which overlapped with the fluid phase WAXS and general background from air, He, and water vapor. Subtracting background scattering data taken at incident angle $-\omega$ removed these unwanted features in the scattering

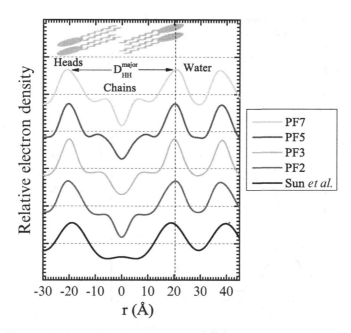

Fig. 3.40 Electron density profiles along Slice A in the major arm, shown in Fig. 3.27, calculated using the phase factors predicted by different fits as indicated in the legend. The distance r is measured from the bilayer center. A cartoon of lipids is shown at the *top*, designating different parts of the profile as the lipid headgroup and chains. The *dashed horizontal lines* show electron density $\rho(r) = 0$. The *dashed vertical line* is to facilitate visual comparison of the headgroup positions

data, resulting in a sample scattering image, Fig. 3.46 (bottom, left panel). This sample scattering image was then transformed to the sample q-space using the relationship between the CCD pixel positions and the sample q-space given by Eqs. (3.6) and (3.7). The nonlinearity of this relationship is not negligible and must be taken into account for wide angle scattering data. The black regions in the sample q-space image, Fig. 3.46 (bottom, right panel), are the regions of q-space that were not probed by the detector when $\omega = 0.5°$. Because of the nonlinearity in the transformation, straight detector edges were turned into curves, the effect of which was most visible near the meridian $q_r = 0$. All nGIWAXS data in this chapter were reduced in the same manner.

Because of chain disordering in the fluid phase, chain-chain scattering gives rise to diffuse scattering that is broad in both the q_r and q_z directions [44]. These fluid phase data were collected with a low resolution setup to maximize intensity. The low resolution (Δq_r 0.032 Å^{-1}) did not pose a problem for analysis of these data because observed features were broad. Fluid phase WAXS is most intense at the equator. However, scattering very near the equator was strongly absorbed by

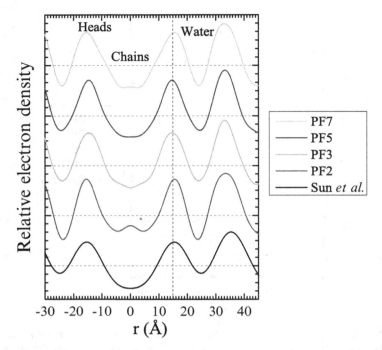

Fig. 3.41 Electron density profiles along Slice B in the minor arm, shown in Fig. 3.27, calculated using the phase factors predicted by different fits as indicated in the legend. The distance r is measured from the bilayer center. The *dashed horizontal lines* show electron density $\rho(r) = 0$. The *dashed vertical line* is to facilitate visual comparison of the headgroup positions

the sample and substrate, so observing the peak in the fluid phase WAXS required the tWAXS experimental geometry in the next section. Figure 3.47 plots intensity along q_r showing that the fluid phase WAXS was centered at $q \approx 1.41\,\text{Å}^{-1}$. This corresponds to an average chain-chain distance of 4.5 Å. A Lorentzian fit to the profile resulted in the full width half maximum (FWHM) $\Delta q_r = 0.288\,\text{Å}^{-1}$.

Figure 3.48 shows nGIWAXS of the the DMPC $L_{\beta I}$ gel phase that occurs at the highest hydration [42, 45], collected with the high resolution setup. Because exposure time was short, the data did not have much intensity, but the (2,0) peak was clearly visible on the equator. When the peak profile of the (2,0) peak in q_r was fitted to a Lorentzian, we obtained an excellent fit with its FWHM $\Delta q_r = 0.014\,\text{Å}^{-1}$, centered at $q_r = 1.479\,\text{Å}^{-1}$. This is the instrumental resolution as discussed in Sect. 3.2.2.3.

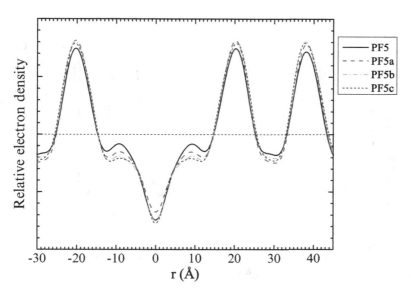

Fig. 3.42 Variation in the electron density profile along Slice A shown in Fig. 3.27. The distance *r* is measured from the bilayer center. The *dashed horizontal line* shows electron density $\rho(r) = 0$. Reversing the sign of the (3, 0) phase in PF5 resulted in the *dashed* profile (PF5a). Reversing the sign of the (1, 3), (3, 0), (6, 0), and (9, 0) resulted in the *dash-dotted* profile (PF5b). Reversing the sign of the (1, 3), (3, 0), (6, 0–4), and (9, 0) resulted in the *short dashed* profile (PF5c)

3.7.2 Ripple Phase

Figure 3.49 shows nGIWAXS from an oriented DMPC film in the ripple phase for $D = 60.8\,\text{Å}$, collected with the high resolution setup. We observed a stronger peak and a weaker one off the equator. The maximum intensity of the stronger peak was at $(q_r, q_z) \approx (1.478 \pm 0.002\,\text{Å}^{-1}, 0.20 \pm 0.01\,\text{Å}^{-1})$ as shown in Fig. 3.50. The weaker peak was observed closer to the equator, and separation of this peak from the stronger one was visible between $q_z = 0.10$ and $0.14\,\text{Å}^{-1}$, indicating that the center of this peak was approximately $(q_r, q_z) \approx (1.457 \pm 0.004\,\text{Å}^{-1}, 0.12 \pm 0.02\,\text{Å}^{-1})$. Because of absorption of X-rays due to the sample, intensity became attenuated as one approaches the equator. Very close to the equator, there is Vineyard-Yoneda peak that is due to constructive interference with scattering from the substrate [46, 47], which we will not consider. Absorption and Vineyard-Yoneda peak did not affect determination of the ripple peak positions as the ripple peaks were located at sufficiently large q_z. The positions of the peaks were consistent with those observed in transmission wide angle scattering, which is discussed in the next section.

We also investigated dependence of the ripple WAXS on the interbilayer *D*-spacing. Figure 3.51 compares nGIWAXS at two different *D*-spacings, showing that chain scattering did not depend on the *D*-spacing in this range. Figure 3.51 (left) also shows a weak feature that looks like an arc coming from the chain peak,

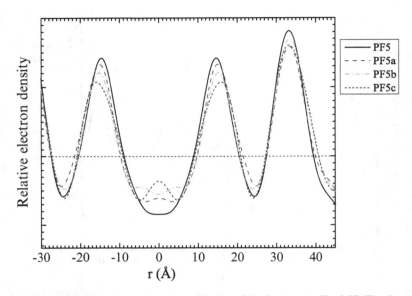

Fig. 3.43 Variation in the electron density profile along Slice B shown in Fig. 3.27. The distance r is measured from the bilayer center. The *dashed horizontal line* shows electron density $\rho(r) = 0$. Reversing the sign of the (3, 0) phase in PF5 resulted in the *dashed* profile (PF5a). Reversing the sign of the (1, 3), (3, 0), (6, 0), and (9, 0) resulted in the *dash-dotted* profile (PF5b). Reversing the sign of the (1, 3), (3, 0), (6, 0–4), and (9, 0) resulted in the *short dashed* profile (PF5c)

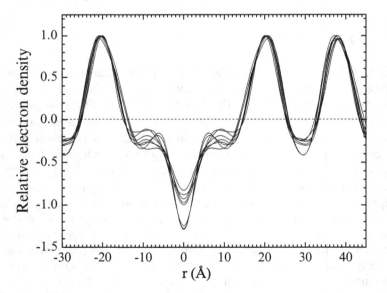

Fig. 3.44 Major arm electron density profiles from PF1, 2, 3, 5, 7, 5a, 5b, and 5c, along Slice A shown in Fig. 3.27. The distance r is measured from the bilayer center. Curves are normalized at the headgroup peaks

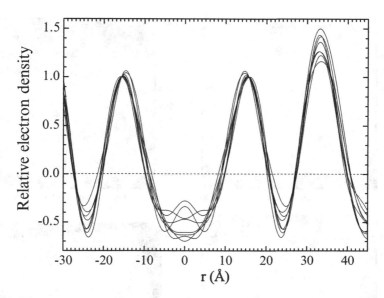

Fig. 3.45 Minor arm electron density profiles from PF1, 2, 3, 5, 7, 5a, 5b, and 5c, along Slice B shown in Fig. 3.27. The distance r is measured from the bilayer center. Curves are scaled to each other at the headgroup peaks

Table 3.18 Estimated structural quantities

A	18.5 ± 0.4 Å
x_{M}	99.2 ± 1.9 Å
ξ_{M}	$10.5° \pm 0.2°$
ξ_{m}	$22.0° \pm 1.0°$
$D_{\mathrm{HH}}^{\mathrm{major}}$	41.2 ± 0.7 Å
$D_{\mathrm{HH}}^{\mathrm{minor}}$	30.2 ± 1.0 Å
D	57.8 Å
λ_r	145.0 Å
γ	$98.2°$

quantified in Fig. 3.52. This feature extended from $\phi = 0°$ to at least $60°$, and is likely to be scattering due to mosaic spread.

We estimated the width of the stronger peak by fitting the intensity profile in q_r to two Lorentzians as shown in Fig. 3.53. The fit resulted in the FWHM $\Delta q_r = 0.025$ Å^{-1} centered at 1.478 Å^{-1} and $\Delta q_r = 0.140$ Å^{-1} centered at 1.464 Å^{-1}. A fit with a single Lorentzian was not very good, and a broader Lorentzian was necessary to produce a reasonable fit. We also fitted the peak profile in q_r at $q_z = 0.12$ Å^{-1}, where two distinct peaks were observed (Fig. 3.54). The two sharp peaks fitted with Lorentzians yielded the FWHM of about 0.025 Å^{-1}, consistent with the FWHM obtained for the stronger peak. The widths and positions of the observed peaks are summarized in Table 3.19.

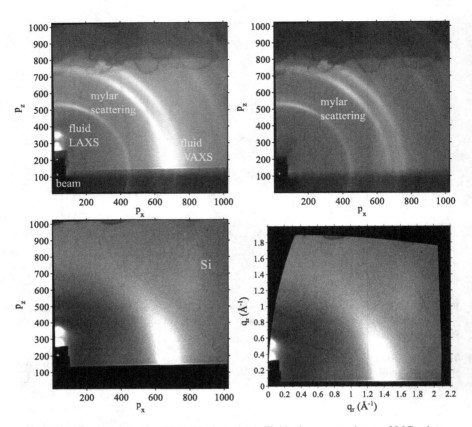

Fig. 3.46 Data reduction of nGIWAXS data. (*top*) Fluid phase scattering at 30 °C taken at $\omega = 0.5°$ (*left*) and at $-0.5°$ (*right*) with the low resolution setup during the 2011 run. The sample width $w_s = 2$ mm. The fluid phase LAXS is also visible near the beam. The *darker region* below the equator defined by the beam vertical position p_z was due to the substrate. The beam was visible through the semitransparent beam stop. Scattering at $p_z > 750$ was the shadow cast by the electrical wires and thermal shielding in the hydration chamber. (*bottom*) The background subtracted image (*left*) and corresponding image in the sample q-space (*right*). Except for some minor leftover scattering, background and mylar scattering was removed. The weak scattering labelled Si also occurred from a bare Si wafer. Because the meridian was not exactly along the vertical pixels, the background subtracted image was rotated by $\sim 1°$ in the clockwise direction before the q-space transformation. The data reduction was done using MATLAB

As Fig. 3.54 shows, the double Lorentzian fit was only successful within a limited range in q_r. This could be due to an underlining broad peak like the one shown in Fig. 3.53. To investigate this possibility, we fitted the same peak profile with three Lorentzians with fixed widths. Two of the Lorentzians had fixed widths of 0.025 Å^{-1} representing the sharp peaks, and the last one had a fixed width of 0.14 Å^{-1} representing the broad peak. Figure 3.54 shows an excellent fit obtained over a large range in q_r, suggesting that the estimated peak widths are not

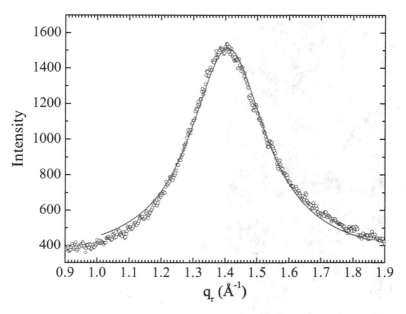

Fig. 3.47 Fluid phase WAXS plotted along q_r at $q_z = 0.012\,\text{Å}^{-1}$. The *solid line* is a Lorentzian fit with its FWHM equal to $0.288\,\text{Å}^{-1}$, centered at $q_r = 1.408\,\text{Å}^{-1}$. Extra intensity at larger q_r was due to water scattering, which led to a slightly asymmetric profile. Resolution was $0.032\,\text{Å}^{-1}$

unreasonable. Curiously, the center of the stronger peak was different at the two different q_z: $(q_r, q_z) = (1.485\,\text{Å}^{-1}, 0.12\,\text{Å}^{-1})$ and $(1.478\,\text{Å}^{-1}, 0.2\,\text{Å}^{-1})$, while the total q was about the same, $\sim 1.49\,\text{Å}^{-1}$.

Figure 3.55 shows the low resolution nGIWAXS of DMPC at $D = 61.0\,\text{Å}$, with the intensity ~ 10 times greater than the high resolution nGIWAXS shown in Fig. 3.49. The low resolution ripple WAXS pattern was similar to the high resolution one, and we did not see any weak feature that could not be observed with the high resolution experiment. Figure 3.56 plots the intensities in q_r at $q_z = 0.012$ and $0.020\,\text{Å}^{-1}$. Due to the low instrumental resolution, the stronger and weaker Bragg rods were not separated; otherwise, the peak profiles were similar to those in the high resolution nGIWAXS shown in Fig. 3.50, indicating that the statistics of the high resolution data were adequate for the ripple WAXS. For completeness, radial intensity profiles $I(q)$ of the low resolution nGIWAXS are plotted in Fig. 3.57.

3.8 tWAXS: Results

Figure 3.58 (left) shows background subtracted transmission wide angle X-ray scattering (tWAXS) of the DMPC gel $L_{\beta I}$ phase obtained at $\omega = 45°$. The background scattering image was collected by replacing the sample with a bare Si wafer. Imperfect subtraction of mylar scattering can be seen in the background

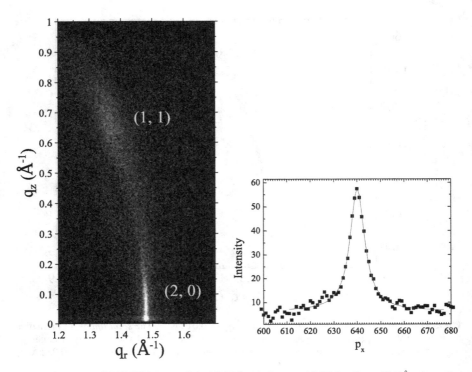

Fig. 3.48 (*Left*) nGIWAXS image of the DMPC gel phase at 10 °C for $D = 57.7$ Å where the sample was in the $L_{\beta I}$ phase. The Miller index for the (2, 0) and (1, 1) Bragg rods are shown. The (2, 0) peak was at $q_r = 1.479$ Å^{-1}, corresponding to $d_{20} = 4.25$ Å. Linear grayscale. (*Right*) The (2, 0) peak plotted along horizontal pixels p_x. The *solid line* is a Lorentzian fit to the data, resulting in the FWHM of ∼8 pixels, corresponding to $\Delta q = 0.014$ Å^{-1}, which is the same as the instrumental resolution estimated in Sect. 3.2.2. The scattering appears weak compared to fluid phase scattering shown in Fig. 3.47 because the high resolution employed here had 100 times smaller intensity

subtracted image. This was most likely due to slight displacement of mylar windows when the sample was replaced with a bare wafer. Three main reflections whose Miller indices are (2, 0), (1, 1), and (1, −1) were observed along with the (1, ±1) satellite peaks. Because the data were taken at $\omega = 45°$, the WAXS pattern appeared on the CCD detector very differently from the respective pattern in the sample q-space. Therefore, the CCD to q-space transformation shown in Fig. 3.58 (right) was important in analyzing the tWAXS data.

Figure 3.59 plots intensity of the $L_{\beta I}$ (2, 0) Bragg rod in q_r shown in Fig. 3.58. A Lorentzian fit to the intensity profile was excellent, with the FWHM $w = 0.020$ Å^{-1}. This value is the same as the instrumental resolution Δq_r estimated in Table 3.4. As was the case for the nGIWAXS (see Sect. 3.7.1), a Gaussian did not fit the data well, indicating that the resolution function for our WAXS experiments could be approximated by a Lorentzian but not Gaussian.

Fig. 3.49 High resolution
nGIWAXS of the DMPC
ripple phase for $D = 60.8$ Å.
The angle of incidence ω was
0.2°. The stronger peak was
at $(q_r, q_z) \approx$
$(1.478 \pm 0.002$ Å^{-1},
0.20 ± 0.01 Å$^{-1})$. The
weaker peak was at $(q_r, q_z) \approx$
$(1.457 \pm 0.004$ Å^{-1},
0.12 ± 0.02 Å$^{-1})$. The
scattered intensity along the
line slightly above $q_z = 0$ Å^{-1}
is Vineyard-Yoneda scattering
involving the substrate
[46, 47]

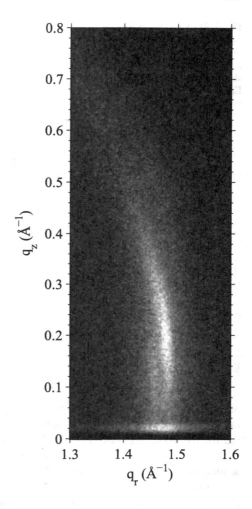

Figure 3.60 shows the tWAXS pattern of the ripple phase after the CCD
to q-space transformation. The stronger peak observed in nGIWAXS was also
observed at approximately the same location. Because of a lower instrumental
resolution than in the nGIWAXS experiment, the weaker peak was not as well
separated from the stronger peak. Figure 3.61 (right) shows a hint of the weak
peak at $q_z = 0.12$ Å^{-1}. This data set taken in the 2011 run motivated the higher
instrumental resolution nGIWAXS experiments in the 2013 run, but it was not
possible to do tWAXS on that run.

The length $L_z = L\cos\theta$ of scattering entities in the bilayer normal z direction
can be estimated by measuring the full length Δq_z of the (2, 0) Bragg rod in q_z
in the $L_{\beta I}$ phase [49], the relation between them being $\Delta q_z = 4\pi/L_z$ from the
$\mathrm{sinc}(q_z L_z/2)$ dependence from simple single slit diffraction. Figure 3.62 (left) shows
intensity of observed Bragg rods along q_z averaged in q_r for the gel and ripple

Fig. 3.50 q_r swaths of the
ripple WAXS, each averaged
over 0.02 Å^{-1} in q_z. The
central q_z values of swaths are
shown in the figure legend.
Each curve is shifted by 100
vertically

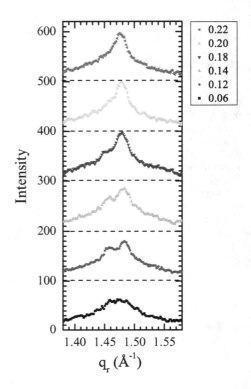

phases. The full length Δq_z for the (2, 0) gel phase peak shown in Fig. 3.62 was
measured to be $\sim$0.4 Å^{-1}, corresponding to $L_z \approx 31$ Å. This value corresponds to
the hydrocarbon thickness $2D_c = 30.2$ Å reported by [42]. This value of L_z, or
$L \approx 37$ Å using $\theta = 32.3°$, indicates that chains in the upper and lower monolayers
scatter coherently, which has been shown to be the case for DMPC [45] and DPPC
[49]. Figure 3.62 (right) compares Δq_z in the ripple and gel phases, showing that
Δq_z was almost the same in both phases. Therefore, chains in the ripple phase are
also coupled between the monolayers. We note that mosaic spread of the sample
would make the apparent Δq_z larger, but negligibly so as the angle subtended by the
Bragg rod is $\tan^{-1}(0.4/1.5) = 15°$, far larger than the mosaicity.

Finally, Fig. 3.63 plots q_z swaths along the weaker Bragg rod and along the entire
ripple WAXS pattern. We found no obvious intensity maxima near the equator,
confirming that the center of the weaker Bragg rod also has non-zero q_z as was
suggested in Sect. 3.7. There was also no sign of a third feature, and deconvolution
of intensity was consistent with two Bragg rods overlapping with their full lengths
given by $\sim$0.4 Å^{-1}.

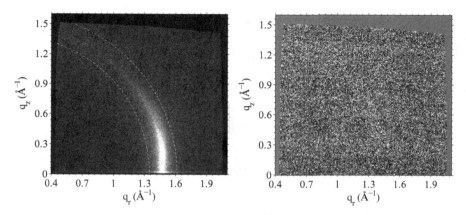

Fig. 3.51 nGIWAXS of the DMPC ripple phase for $D = 59.2\,Å$ (*left*) and difference between $D = 59.2$ and $60.8\,Å$ (*right*). The difference shows no obvious feature, indicating that the ripple WAXS patterns at the two D-spacings were identical within error. The angle of incidence ω was $0.2°$. The data were taken with the high resolution setup. The color is linear grayscale. The *dashed lines* are a guide to the eye

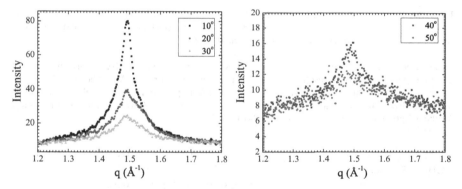

Fig. 3.52 Radial intensity profiles for various angles ϕ. ϕ is an angle measured from the equator (see Fig. 3.17) in q-space. Intensities are plotted as a function of total q, along constant ϕ indicated in the figure legend. Intensities were averaged over $\pm 5°$ to improve statistics

3.9 Thin Rod Model

We attempt to understand the WAXS data by considering models for the chain packing. As the prevailing hypothesis is that the major arm of the ripple is like a gel phase, we begin by reviewing gel phase models in the next subsection. We then consider the scattering consequences of tilting these models out of plane by the angle $\xi_M \approx 10.5°$ obtained from the LAXS analysis.

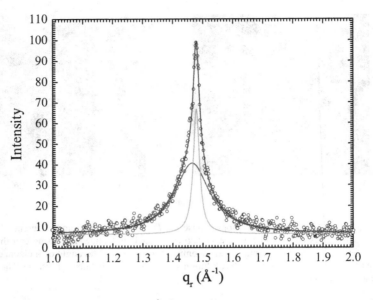

Fig. 3.53 Peak profile in q_r at $q_z = 0.2\,\text{Å}^{-1}$ fitted to the sum of two Lorentzians. The FWHM and center obtained were $0.025\,\text{Å}^{-1}$ and $1.478\,\text{Å}^{-1}$ (*narrow peak*) and $0.140\,\text{Å}^{-1}$ and $1.464\,\text{Å}^{-1}$ (*wide peak*), respectively. A sum of the two Lorentzian fits is also shown

Table 3.19 Wide angle peak positions and widths for the ripple phase with comparison to the gel and fluid phases. The only resolution limited width was for gel phase Δq_r. The tWAXS entries are from Sect. 3.8. The q value of gel $(\pm 1, 1)$ is taken from [42]. For the ripple phase the ratio of integrated Bragg rod intensities was $R_{ripple} = I_{weak}/I_{strong} \approx 0.54\text{--}0.67$ in comparison to a ratio for the gel phase of $R_{gel} = I_{(2,0)}/I_{(1,1)} \approx 0.8\text{--}1.0$ [48]

Type	Peaks	q (Å^{-1})	q_r (Å^{-1})	q_z (Å^{-1})	Δq_r (Å^{-1})	Δq_z (Å^{-1})
nGIWAXS	Stronger	1.491	1.478 ± 0.002	0.20 ± 0.01	0.025	
	Weaker	1.462	1.457 ± 0.004	0.12 ± 0.02	0.025	
	Broader	1.463–1.478	1.458–1.464	0.12–0.20	0.140	
	gel (2, 0)	1.479	1.479	0	0.014	
	gel (± 1, 1)	1.536	1.357	0.720		
	Fluid	1.41			0.288	
tWAXS	Stronger	1.496	1.484 ± 0.002	0.20 ± 0.01	0.028	0.4
	Weaker	1.466	1.461 ± 0.004	0.12 ± 0.02	0.036	
	Broader	1.478–1.488	1.473–1.475	0.12–0.20	0.088, 0.213	
	gel (2, 0)	1.479	1.479	0	0.020	0.4

3.9.1 Gel Phase Model

The fully hydrated gel phase of DMPC consists of hydrocarbon chains that are basically straight and cooperatively tilted by an angle θ from the bilayer normal

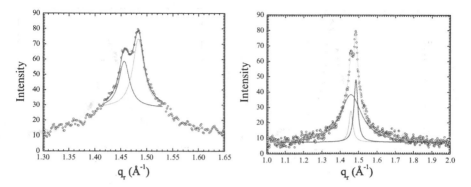

Fig. 3.54 (*Left*) Peak profile in q_r at $q_z = 0.12\,\text{Å}^{-1}$ fitted to the sum of two Lorentzian functions. The FWHM and center obtained were $0.025\,\text{Å}^{-1}$ and $1.457\,\text{Å}^{-1}$ (*left peak*) and $0.026\,\text{Å}^{-1}$ and $1.484\,\text{Å}^{-1}$ (*right peak*), respectively. The fit was limited within a range in which fits were reasonable. (*Right*) The same peak profile fitted to the sum of three Lorentzians. The FWHM were constrained to $0.025\,\text{Å}^{-1}$ (*right, narrow peak*), $0.025\,\text{Å}^{-1}$ (*left, narrow peak*), and $0.14\,\text{Å}^{-1}$ (*wide peak*). The centers were found to be $1.485\,\text{Å}^{-1}$ (*right, narrow peak*), $1.458\,\text{Å}^{-1}$ (*left, narrow peak*), and $1.458\,\text{Å}^{-1}$ (*wide peak*)

Fig. 3.55 Low resolution nGIWAXS of the DMPC ripple phase for $D = 61.0\,\text{Å}$. The angle of incidence ω was $0.2°$. The instrumental resolution $\Delta q_r = 0.032\,\text{Å}^{-1}$

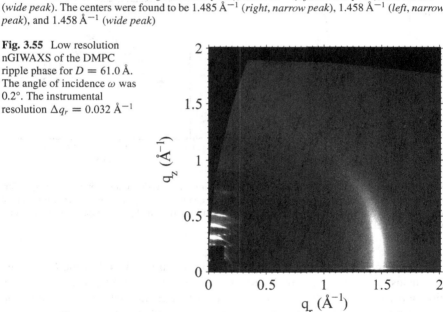

[42, 45, 49, 50]. This is called the $L_{\beta I}$ phase in which each chain is tilted toward a nearest neighbor chain. At lower hydration the chains tilt differently. We will also focus on the $L_{\beta F}$ phase in this section. The chains will be modeled as thin rods. The basic geometry of the $L_{\beta I}$ and $L_{\beta F}$ phases is shown in Fig. 3.64. Reference [50] emphasized that the chains are tilted in the same direction in both monolayers. It also allowed for translational offsets that we will set to zero for simplicity.

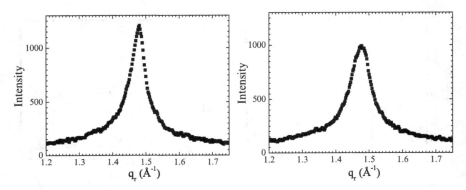

Fig. 3.56 q_r swaths along $q_z = 0.020$ (*left*) and 0.012 Å^{-1} (*right*) of the low resolution nGIWAXS data shown in Fig. 3.55. The *left* profile is along the stronger Bragg rod observed in the high resolution nGIWAXS shown in Fig. 3.49. Due to the low instrumental resolution $\Delta q_r = 0.032$ Å^{-1}, the weaker Bragg rod was not separated from the stronger one

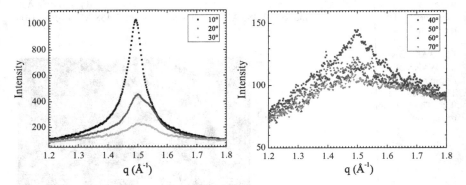

Fig. 3.57 Radial intensity profiles of the low resolution nGIWAXS data shown in Fig. 3.55 for various angles ϕ. ϕ is an angle measured from the equator (see Fig. 3.17) in q-space. Intensities are plotted as a function of total q, along constant ϕ indicated in the figure legend. Intensities were averaged over $\pm 5°$ to improve statistics

The unit cell customarily employed is indicated in Fig. 3.64. For the $L_{\beta I}$ phase, the chains are tilted along the **b** direction as shown in Fig. 3.64 and along the **a** direction for the $L_{\beta F}$ phase. It may be noted that chain packing in a plane that is perpendicular to the chains (and therefore not parallel to the bilayer) is nearly hexagonal; if the packing were hexagonal and if the chains had zero tilt, then in Fig. 3.1c one would have $b = a/\sqrt{3}$, which becomes $b = a/(\cos\theta\sqrt{3})$ with tilt. The Laue conditions for allowed reflections are

$$q_x = \frac{2\pi m}{a} \tag{3.51}$$

and

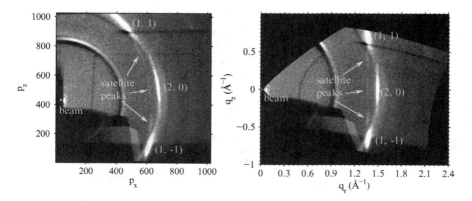

Fig. 3.58 Transmission WAXS at fixed angle $\omega = 45°$ of the DMPC gel $L_{\beta I}$ phase observed on the CCD detector (*left*) and its corresponding pattern in the sample q-space (*right*). Bragg rods were indexed as (2, 0), (1, 1) and (1, −1). The satellite peaks of (1, ±1) reflections were also labeled. The *black region* in the *right image* corresponds to inaccesible q-space at $\omega = 45°$. The edges of the sample q-space image were distorted due to the nonlinear relation between the detector pixels and the sample q-space as discussed in Sect. 3.7. A ring of intensity at $q \approx 0.9 \, \text{Å}^{-1}$ is due to imperfect subtraction of the mylar scattering. Residual mylar scattering is also visible near and at slightly larger q than the (2, 0) Bragg rod

$$q_y = \frac{2\pi n}{b}, \tag{3.52}$$

where m and n are integers. Equations (3.51) and (3.52) establish the location of possible lines of scattering (Bragg rods). The modulation of the intensity along these rods is derived from the square of the unit cell form factor

$$F(\mathbf{q}) = \int_0^a dx \int_0^b dy \int_{-\frac{L}{2}\cos\theta}^{\frac{L}{2}\cos\theta} dz \, \rho(\mathbf{r}) \exp(i\mathbf{q} \cdot \mathbf{r}). \tag{3.53}$$

Our thin rods are modeled as delta functions

$$\rho(\mathbf{r}) = \delta(x - \alpha z, y - \beta z) + \delta(x - a/2 - \alpha z, y - b/2 - \beta z) \tag{3.54}$$

where for the general case that the chain tilt is oriented at angle ϕ relative to the x axis

$$\alpha = \tan\theta \cos\phi \tag{3.55}$$

and

$$\beta = \tan\theta \sin\phi. \tag{3.56}$$

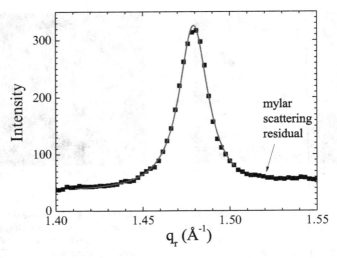

Fig. 3.59 q_r swath of the (2, 0) Bragg rod at $q_z = 0$ Å^{-1}. The *solid line* is a Lorentzian fit, with the center at $q_r = 1.479$ Å^{-1} and FWHM $w = 0.020$ Å^{-1}, which is the same as the instrumental resolution estimated in Table 3.4. Additional intensity in the right tail of the peak is due to imperfect subtraction of the mylar scattering. Intensities are averaged between $q_z = -0.05$ and 0.05 Å^{-1}

For the $L_{\beta I}$ phase, $\phi = \pi/2$ and for the $L_{\beta F}$ phase, $\phi = 0$. Continuing with the general ϕ case for awhile, defining $\gamma = \alpha q_x + \beta q_y + q_z$ yields

$$F(\mathbf{q}) = \int_{-\frac{L}{2}\cos\theta}^{\frac{L}{2}\cos\theta} dz\, \rho(\mathbf{r}) e^{i\gamma z} (1 + e^{\frac{q_x a}{2} + \frac{q_y b}{2}}). \tag{3.57}$$

The phase factor $1 + e^{q_x a/2 + q_y b/2}$ vanishes unless the sum $m + n$ of the Laue indices (mn) is even. Only the lowest orders $(\pm 2, 0)$ and $(\pm 1, \pm 1)$ have observable intensity. For the simple thin rod model in Eq. (3.54)

$$F(q_z) = \frac{4}{\gamma} \sin\left(\frac{\gamma L \cos\theta}{2}\right) \tag{3.58}$$

so the intensity $|F(q_z)|^2$ is modulated along each Bragg rod and maximum intensity occurs when $\gamma = 0$ which, upon reversing the convention for the sign of q_z, means that the wide angle peaks are centered at

$$q_z^{mn} = \alpha q_x + \beta q_y = \alpha \frac{2\pi m}{a} + \beta \frac{2\pi n}{b}. \tag{3.59}$$

For the $L_{\beta I}$ phase with $\phi = \pi/2$, one has

Fig. 3.60 tWAXS image of the DMPC ripple phase at 18 °C and $D = 60.3$ Å. The instrumental resolution $\Delta q_r = 0.020$ Å^{-1}, as estimated in Sect. 3.2.2.3 and also from the $L_{\beta I}$ (2, 0) width in Fig. 3.59. Color is linear grayscale with *white* being most intense and *black* being least intense

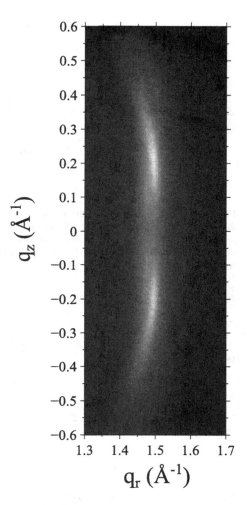

$$0 = q_{z\beta I}^{20} = q_{z\beta I}^{-20} \tag{3.60}$$

$$\frac{2\pi}{b} \tan \theta = q_{z\beta I}^{11} = q_{z\beta I}^{-11} = -q_{z\beta I}^{1-1} = -q_{z\beta I}^{-1-1} \tag{3.61}$$

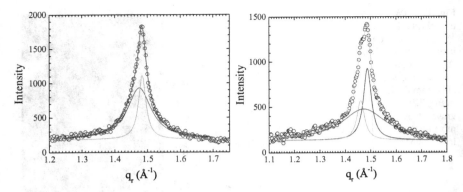

Fig. 3.61 (*Left*) Peak profile of the ripple tWAXS in q_r at $q_z = 0.2$ Å^{-1}. The *line* is a fit to the sum of two Lorentzians. The FWHM and center obtained were 0.028 and 1.484 (*narrow peak*) and 0.088 and 1.475 Å^{-1} (*wide peak*), respectively. (*Right*) Corresponding peak profile at $q_z = 0.12$ Å^{-1}. The *line* is a fit to the sum of three Lorentzians. The FWHM and center obtained were 0.038 and 1.488 (*right, narrow peak*), 0.036 and 1.461 (*left, narrow peak*), and 0.213 and 1.473 Å^{-1} (*wide peak*), respectively

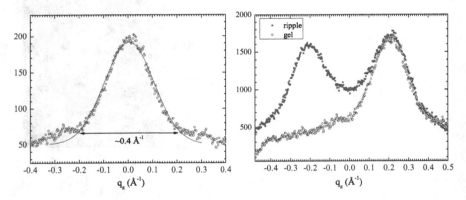

Fig. 3.62 (*Left*) q_z swath of the gel (2,0) Bragg rod. The *solid line* is a Gaussian fit with the FWHM of 0.23 Å^{-1}. (*Right*) q_z swath of the ripple stronger Bragg rod averaged between 1.481 Å^{-1} and 1.510 Å^{-1} in q_r (*solid squares*) and the gel (2,0) peak scaled and shifted in q_z to guide visual comparison (*open black circles*). The instrumental resolution $\Delta q_z = 0.0074$ Å^{-1}

For the $L_{\beta F}$ phase with $\phi = 0$

$$\frac{4\pi}{a} \tan \theta = q_{z\beta F}^{20} = -q_{z\beta F}^{-20} \tag{3.62}$$

and

$$\frac{2\pi}{a} \tan \theta = q_{z\beta F}^{11} = q_{z\beta F}^{1-1} = -q_{z\beta F}^{-11} = -q_{z\beta F}^{-1-1} \tag{3.63}$$

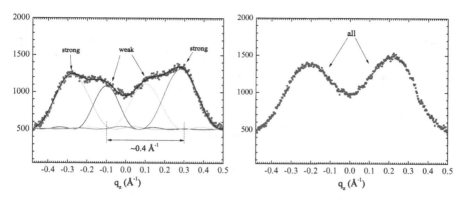

Fig. 3.63 q_z swath averaged between 1.465 and 1.481 Å^{-1} (*left*) and between 1.465 and 1.51 Å^{-1} (*right*) in q_r. The *left plot* is approximately the q_z profile along the weaker peak while the *right* profile extends over the entire ripple WAXS pattern. Intensity on the *left* was fit to four sinc functions, two above and two below the equator. The widths of sinc functions shown by the *arrow* were ∼0.4 Å^{-1}, consistent with the gel $L_{\beta I}$ (2, 0) Bragg rod (Fig. 3.62)

One can verify, using these equations and the Laue equations for q_x and q_y that the magnitudes $q^{\pm 20}$ and $q^{\pm 1\pm 1}$ of the total scattering vectors are equal when the packing of the chains is hexagonal in the tilted chain plane.

In q-space the powder averaged gel phase pattern consists of circles in q_x and q_y centered on $q_x = 0 = q_y$ and with the values of q_z given in Eqs. (3.60)–(3.63). The location of observed scattering in lab space $\mathbf{k}$ is obtained using the Ewald sphere, centered at $\mathbf{k} = 0$ with radius $2\pi/\lambda$ and with the $\mathbf{q} = 0$ center of the q-space pattern located at $\mathbf{k} = (0, |\mathbf{k}|, 0)$. The q-space pattern is tilted by the angle ω when the sample is tilted relative to the laboratory frame; for grazing incidence, the q_z and k_z axes are parallel and offset by $2\pi/\lambda$ in the k_y beam direction. The direction of scattering for the powder averaged gel phase is given by the laboratory $\mathbf{k}$ values where the q-space pattern intersects the Ewald sphere. Each of the (mn) rings generally intersects twice with opposite signs for k_x corresponding to opposite sides of the meridian on the CCD. The only rings that give obervable scattering in the gel phase are the (± 20) and the ($\pm 1 \pm 1$) rings. However, some of these six rings may coincide. For the $L_{\beta I}$ phase (± 20), (± 11) and ($\pm 1 - 1$) are pairwise identical, so there are three primary reflections on each side of the meridian. For the $L_{\beta F}$ phase (1 ± 1) and (-1 ± 1) are pairwise identical, so there are four primary peaks on each side of the meridian.

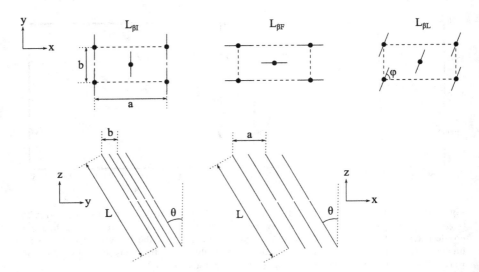

Fig. 3.64 Lattice and geometry of thin rod model. The chains are represented as *solid lines*. The unit cell is drawn in the *dashed lines*. *Top views* of $L_{\beta I}$, $L_{\beta F}$, and $L_{\beta L}$ phases (*top*) and *side views* of $L_{\beta I}$ and $L_{\beta F}$ (*bottom*) are shown. **a** and **b** are unit cell vectors, and $a > b$. ϕ is an in-plane azimuthal angle. θ is the chain tilt angle with respect to the bilayer normal z. Chains are tilted toward the nearest neighbor in the $L_{\beta I}$ phase with $\phi = \pi/2$. The $L_{\beta I}$ phase is observed in the fully hydrated gel phase of DMPC. In the $L_{\beta F}$ phase, the chains are tilted toward the next nearest neightbor ($\phi = 0$)

3.9.2 Ripple Model

A reasonable hypothesis is that the major arm of the ripple has similar internal structure to a gel phase, with the major difference that the plane of the major arm is tilted relative to the substrate. That suggests that the predicted ripple pattern might be the same as would be obtained by tilting the in-plane powder averaged gel phase. However, this would be a fundamental error because the operations of tilting and in-plane powder averaging do not commute. It is necessary first to tilt the gel phase q-space pattern and then to powder average it about the laboratory k_z axis.

Furthermore, the axis for tilting matters, so it is important to define all angles carefully as shown in Fig. 3.65. We continue to define the chain tilt angle relative to the bilayer normal by θ. The tilt of the major arm will be defined by a rotation angle ξ about an axis in the (x, y) plane and the angle that this axis makes with the x axis will be defined to be ζ. Starting from the q values obtained for the various gel phases, the proper order of rotations is first to rotate the orientation of the lattice with respect to the lab frame; this involves the standard rotation of the (x, y) plane about the z axis by angle ζ. Then, the gel phase is rotated about the new in-plane x axis. The rotated q value will be denoted $\tilde{q}$ which has components

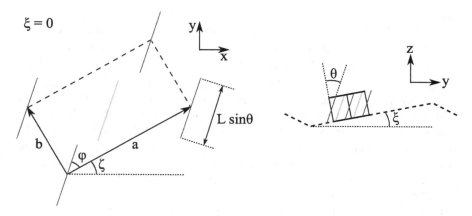

Fig. 3.65 (*Left*) Projection of the unit cell (a, b) on the laboratory (x, y) plane. The unit cell is rotated by ζ compared to a being parallel to x and the direction of chain tilt is rotated by ϕ from the a axis. (*Right*) Without loss of generality, the ripple direction is shown along the y axis and the major side is tilted by ξ. Supposing that the chains are tilted in the y direction only, then the corresponding gel phase could be any $L_{\beta L}$ phase constrained only by $\phi + \zeta = \pi/2$, including the special $L_{\beta I}$ phase with $\phi = \pi/2$ and $\zeta = 0$ and the special $L_{\beta F}$ phase with $\phi = 0$ and $\zeta = \pi/2$

$$\tilde{q}_z^{mn} = q_z^{mn} \cos \xi + q_x^{mn} \sin \xi \sin\zeta - q_y^{mn} \sin \xi \cos \zeta, \tag{3.64}$$

$$\tilde{q}_x^{mn} = q_x^{mn} \cos \zeta + q_y^{mn} \sin \zeta, \tag{3.65}$$

and

$$\tilde{q}_y^{mn} = q_y^{mn} \cos \xi \cos \zeta - q_x^{mn} \cos \xi \sin \zeta + q_z^{mn} \sin \xi. \tag{3.66}$$

As there are many domains in each x-ray exposure, the next step powder averages each (mn) reflection by rotating about the z axis from 0 to 2π. As for the gel phase, the ensuing q space pattern consists of circles parallel to the (x, y) plane with center at $(0, 0, q_z^{mn})$. As noted above for the gel phase, this pattern is tilted by ω when the substrate is tilted for our transmission experiments. Intersections of these circles with the Ewald sphere determines the angle of scattering in the laboratory from which, by standard equations (Sect. 3.2.6), the $\mathbf{q}_{mn}$ are determined.

The most pertinent component is $\tilde{q}_z^{mn}$ as this primarily determines how far reflections are from the meridian. As there are many variable angles, let us consider $\tilde{q}_z^{mn}$ for the most pertinent special cases. It is appropriate here to consider only $\omega = 0$ because experimental data with $\omega \neq 0$ are easily converted to this standard orientation. We will focus on four special cases. First, consider the in-plane orientation ζ of the lattice to have either the longer a axis parallel ($\zeta = 0$) or perpendicular ($\zeta = \pi/2$) to the ripple direction. It may be noted that these two special directions allow uniform packing of the unit cells along the finite ripple direction, whereas the edges of the unit cells are ragged at the boundaries of the major arm for other values of ζ. Also, these two directions are symmetrical.

However, as the lipid molecules are chiral and as there is likely disorder at the boundaries of the major arm, one cannot eliminate general ζ angles a priori. We will also focus on the special orientations of the tilt direction that correspond to the $L_{\beta I}$ gel phase ($\phi = \pi/2$), which we will henceforth call $P_{\beta I}^{\zeta}$ phases, and the $L_{\beta F}$ gel phase ($\phi = 0$), to be called $P_{\beta F}^{\zeta}$ phases, recognizing, of course, that we are only modeling the major arm of the $P_{\beta'}$ ripple phase. It will also be convenient to simplify to hexagonal packing of the hydrocarbon chains as the orthorhombic symmetry breaking that makes $q_{\text{total}}^{20} \neq q_{\text{total}}^{11}$ is small; then, $b = a/(\sqrt{3}\cos\theta)$ for the $P_{\beta I}^{\zeta}$ phases and $b = a\cos\theta/\sqrt{3}$ for the $P_{\beta F}^{\zeta}$ phases. These simplifications allow us to focus on the chain tilt angle θ and the tilt ξ of the major side for four cases of (ϕ, ζ) and the observable orders $(\pm 2, 0)$ and $(\pm 1, \pm 1)$. The following table shows the values of q_z^{mn}, all divided by $2\pi/a$.

Importantly, tilting the gel phase to form putative ripple major arms breaks the degeneracy of many of the gel phase rings. Most notably, all the degeneracies are broken in the $P_{\beta I}^{\zeta=\pi/2}$ special case whereas none are broken in $P_{\beta F}^{\zeta=\pi/2}$. The magnitude of the q_z symmetry breaking is typically $(4\pi/a)\sin\xi \approx 0.32$ Å^{-1} for $\xi = 10.5°$. As $\Delta q_z \approx 0.4$ Å, broken symmetry Bragg rods would be predicted to overlap considerably. This could blur them into apparently single Bragg rods, but with larger Δq_z than the intrinsic value of each Bragg rod.

3.10 Combining WAXS and LAXS Results for the Major Arm

Table 3.21 lists our new WAXS results for the ripple phase and some older results for the gel phase [42]. The definitions of the structural quantities in Table 3.21 are illustrated in Fig. 3.66. As for the well hydrated $L_{\beta I}$ gel phase, the total q for independent reflections is different due to orthorhombic symmetry breaking of hexagonal chain packing, but the difference is smaller for our ripple data, so we have ignored this in our models in Sect. 3.9.2. Our experimental result for Δq_z in Sect. 3.8 for the two ripple Bragg rods corresponds to scattering units with length L_z equal to the hydrocarbon thickness that is often called $2D_c$ (see Fig. 3.66). Overlap of Bragg

Table 3.20 q_z^{hk} divided by $2\pi/a$

	$(\pm 2, 0)$	$(\pm 1, 1)$	$(\pm 1, -1)$
$L_{\beta I}$	0	$\sqrt{3}\sin\theta$	$-\sqrt{3}\sin\theta$
$P_{\beta I}^{\zeta=0}$	0	$\sqrt{3}\sin(\theta-\xi)$	$-\sqrt{3}\sin(\theta-\xi)$
$P_{\beta I}^{\zeta=\pi/2}$	$\pm 2\sin\xi$	$\sqrt{3}\sin\theta\cos\xi \pm \sin\xi$	$-\sqrt{3}\sin\theta\cos\xi \pm \sin\xi$
	$(\pm 2, 0)$	$(1, \pm 1)$	$(-1, \pm 1)$
$L_{\beta F}$	$\pm 2\tan\theta$	$\tan\theta$	$-\tan\theta$
$P_{\beta F}^{\zeta=0}$	$\pm 2\tan\theta\cos\xi$	$\tan\theta\cos\xi \mp \sqrt{3}\sin\xi/\cos\theta$	$-(\tan\theta\cos\xi \mp \sqrt{3}\sin\xi/\cos\theta)$
$P_{\beta F}^{\zeta=\pi/2}$	$\pm 2(\tan\theta\cos\xi + \sin\xi)$	$\tan\theta\cos\xi + \sin\xi$	$-(\tan\theta\cos\xi + \sin\xi)$

Table 3.21 Quantities for chain packing structure in the major arm. The values for $L_{\beta I}$ and $L_{\beta F}$ are from [42]. $\theta = 18.4°$ and $a = 8.88$ Å were obtained from the measured Bragg rod positions in the $P_{\beta F}^{\zeta=\pi/2}$ phase. d_{cc} is the chain-chain distance

	(m, n)	q (Å^{-1})	q_z (Å^{-1})	θ	d_{cc} (Å)	$2D_c$ (Å)	ℓ_{ch} (Å)
$L_{\beta I}$	(2, 0)	1.479	0	32.3°	4.67	30.1	17.8
	(1, 1)	1.536	0.720				
$L_{\beta F}$	–	–	–	20.4°	–	33.4	17.8
$P_{\beta F}^{\zeta=\pi/2}$	Strong	1.491	0.205	18.4°	4.86	31.2	16.5
	weak	1.460	0.102				

rods required by the symmetry breaking that occurs in three of the four models described in Sect. 3.9.2 would require that each Bragg rod arise from a coherent scattering length even larger than a bilayer. As that is an unorthodox possibility, let us focus on the only case in Table 3.20 that does not have symmetry breaking, namely the $P_{\beta F}^{\zeta=\pi/2}$ phase. For this phase q_z^{20} is twice as large as for $q_z^{\pm 11}$, which is consistent with the ratio of the experimental peak positions, $q_z = 0.2 \pm 0.01$ and 0.12 ± 0.02. Using our LAXS result $\xi_M = 10.5°$ from Table 3.18 and the experimental values in Table 3.19 gives $\theta = 18.4°$. This is considerably smaller than for the well hydrated $L_{\beta I}$ phase. Table 3.21 also has an entry for the $L_{\beta F}$ gel phase that was obtained from Fig. 6 in [42] for a sample that had been partially dehydrated so that the lamellar D spacing was 7 Å smaller and the tilt angle had decreased to 20°. The ripple samples that we have focused on also have D spacings about 7 Å smaller than full hydration ($D_{FH} \approx 66.0$ Å [51]). We therefore advance the hypothesis that our ripple samples have major arms that are like the $L_{\beta F}$ gel phase rather than like the $L_{\beta I}$ gel phase.

A further test of this hypothesis is that the thickness of the major arm is consistent with our LAXS result that gave a head-head thickness $D_{HH} = 41.2$ Å from Table 3.18. In the gel phase, the distance from the peak in the electron density profile to the Gibbs dividing surface for the hydrocarbon core has been found to be about 5 Å[42], so we estimate the ripple hydrocarbon thickness $2D_c$ to be 31.2 Å, a bit larger than the gel phase value. However, the smaller tilt angle implies that the chain length ℓ_{ch} is smaller, using $\ell_{ch} = D_c/\cos\theta$ with values shown in Table 3.21. This difference is reasonable as the ripple phase occurs at higher temperature and more rotameric defects would be expected to occur. Indeed, one g^+tg^- kink per chain would reduce the chain length by 1.27 Å [52, 53] which could account for the difference in ℓ_{ch} in Table 3.21 between the gel and ripple phases. Chain shortening disorder is also consistent with the larger distance between neighboring ripple chains d_{cc} in Table 3.21. More disorder is also consistent with the shorter chain-chain correlation length implied by the larger Δq_r in Table 3.19.

While the positions of the observed peaks are consistent with the major arm of the ripple being a slightly disordered and tilted $L_{\beta F}$ structure, the relative intensities

Fig. 3.66 Definitions of
structural quantities
appearing in Table 3.21.
Chains are tilted with respect
to the bilayer normal z
direction by θ. D_{HH} is the
distance between the maxima
in an electron density profile.
The distance between
adjacent CH_2 groups is
1.27 Å measured along the
chain tilt direction.
$\ell_{ch} = 14 \times (1.27\,\text{Å})$ in the
gel phase. $D_c = \ell_{ch} \cos\theta$

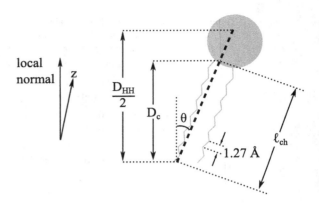

in those Bragg rods do not easily conform. In the $L_{\beta F}$ structure the Bragg rod with
peak at smaller q_z is a superposition of the $(1, 1)$ and $(-1, 1)$ Bragg rods whereas
the rod with peak at larger q_z arises only from the $(2, 0)$ rod. When the chains are
modeled as thin rods, the relative intensities of $(2, 0)$, $(1, 1)$ and $(-1, 1)$ would all be
equal, giving a factor of 2 in the ratio of observed intensities, but the experimental
relative intensity ratio is at most 2/3. Of course, all-trans chains are not thin rods but
the carbons zigzag in a plane, and there have been recent calculations that suggest
that these intensity ratios can vary considerably depending on yet another angle, the
azimuthal rotation of the chains about their long axis [48, 54].

3.11 Discussion

3.11.1 Major Arm

There is significant evidence suggesting that the major arm is like the gel phase
[1, 8, 27, 33]. Figure 3.67 compares our electron density profile in the major arm to
the DMPC gel phase profile reported by Tristram-Nagle et al. [42]. It shows that the
density profile of the major arm is similar to that of gel phase, and the thickness
is comparable between the two phases although the ripple profile does not show
distinction between the phosphate and carbonyl-glycerol headgroups as in the gel
phase. Also, the terminal methyl trough appears to be wider in the ripple major arm,
which could be a sign that the terminal methyl is more disordered in the ripple phase
than in the gel phase. As discussed in Sect. 3.6, however, small features in the ripple
profile are not reliable because they depend on which phase factors were used to
calculate the electron density profile.

tWAXS results further emphasize the gel-like nature of the major arm.
Figure 3.62 shows that the lengths of the Bragg rods in the ripple and gel phases
are approximately the same, indicating that chains in different leaflets in the ripple
major arm also scatter coherently.

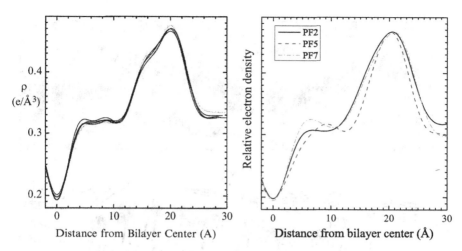

Fig. 3.67 Comparison of the electron density profiles of the DMPC gel phase (*left*) and the major arm in the ripple phase (*right*). The *left figure* is taken from [42]. The ripple major arm profiles were calculated using the phase factors obtained from Fit2, 5, and 7 (*solid*, *dashed*, and *dash-dotted*, respectively). The ripple profiles are scaled vertically to match with the gel phase profile

With the thin rod model developed in Sect. 3.9.2, we indexed the stronger Bragg rod as (2, 0) and weaker one as (1, ±1). The chain tilt angle $\theta = 18.4°$ (see Table 3.21) is consistent with the gel $L_{\beta F}$ phase, supporting our interpretation that chains are in the $P_{\beta F}$ phase. With the result from the tWAXS analysis that chains in opposing leaflets scatter coherently, Fig. 3.68 (top) sketches our proposed chain packing in the major arm, where chains are tilted in the xz-plane with respect to the bilayer normal. This interpretation of the WAXS data is consistent with Fig. 3.67, which shows that the major arm thickness is comparable to the gel phase bilayer thickness, indicating that chains in the ripple major arm are also tilted with respect to the bilayer normal.

On the other hand, Sengupta et al. proposed that major arm chains are oriented nearly parallel to the bilayer normal [11, 12], see Fig. 3.68 (bottom). This chain orientation would lead to a bilayer thickness $D_{HH} = 44.9$ Å with all-trans chains, somewhat thicker than our major arm of $D_{HH} = 41.2$ Å. However, we have argued in Sect. 3.10 that the chains must be slightly disordered for our model; applying a similar argument to the model in [11, 12] would reduce their predicted bilayer thickness to $D_{HH} = 42.4$ Å sufficiently close to our measured value that thickness can not be used to rule out their model. More importantly, the Bragg rod positions based on the chain orientation proposed by Sengupta et al. [11, 12], are inconsistent with the measured Bragg rod positions in our nGIWAXS data. Sengupta et al. proposed the chain packing based on their analysis of the LAXS data from the unoriented sample [6]: specifically, the best fit value of the model parameter ψ, which allows rotation of the transbilayer electron density profile with respect to the stacking z direction (see Sect. 3.5.2) [11, 12]. However, as we emphasized in

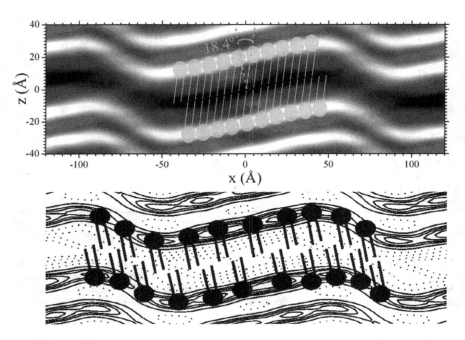

Fig. 3.68 (*Top*) Chain packing in the major arm proposed in this thesis. Chains are drawn as *solid lines*, and headgroups as *circles*. Lipids are overlaid on the electron density map shown in Fig. 3.27. Chains are tilted by 18.4° with respect to the bilayer local normal. Chains in the *upper* and *lower* leaflets are coupled and scatter coherently. (*bottom*) Corresponding chain packing proposed by Sengupta et al., reproduced from [12]. Contours drawn are *contour lines* of the electron density map. Chains are nearly parallel to the bilayer normal of the major arm

Sect. 3.5.4, a model should only be used to obtain the phase factors, and the model parameter values should not be used to infer the actual bilayer structure. Indeed, we also obtained the best fit ψ value similar to the one obtained in [11, 12] (Table 3.12), and naive interpretation of the ψ value would lead to the same chain tilt proposed in [11, 12]. However, the nGIWAXS analysis led to a quite different chain tilt angle. This problem of inferring the actual structure from the model parameters was also evidenced by the values of the ripple amplitude A measured from the calculated electron density maps in Table 3.17, which were more than 3 Å smaller than the model parameter A obtained from the best fits in Table 3.12. Thus, WAXS is essential for determining molecular organization. The combination of LAXS and WAXS analyses yields more structural information than either technique alone.

Our proposed chain packing shown in Fig. 3.68 (top) compares favorably with the chain packing in the major arm predicted by a Landau-Ginzburg theory developed by Kamal et al. that allows a coupling of the lipid tilt field to the chain conformation field [32]. Interestingly, this theory paper [32] states that its results agree with the chain packing reported in [12]. However, [12] showed the chains in the major arm aligned along the local normal as reproduced in Fig. 3.68, in disagreement with

Fig. 5 in [32] which agrees with our result in Fig. 3.68. The chains in the minor arm in [32] are portrayed as disordered fluid-like, also in disagreement with [12]. We discuss the chain packing in the minor arm in the next subsection. In view of the successful prediction of [32] of our new result for the major arm, this theory should provide insights into what causes the ripple phase. Unfortunately, there are 18 parameters in this Landau-Ginzburg theory. While many of the terms are understandable, others are difficult to interpret in terms of interactions between lipid molecules.

The chain-chain correlation length can be estimated by using the Scherrer equation [37],

$$B = \frac{0.94\lambda}{L\cos\theta},$$

where B is the observed FWHM of a Bragg peak, λ is the wavelength, L is the length over which chains are positionally correlated, and θ is the Bragg angle. For the (2, 0) Bragg peak in the gel $L_{\beta I}$ phase, we obtained the FWHM $\Delta q = 0.014\,\text{Å}^{-1}$ and the position of the peak $q = 1.479\,\text{Å}^{-1}$. For our X-ray wavelength $\lambda = 1.175\,\text{Å}$, the Scherrer equation yields $L = 426\,\text{Å}$. Because the width of the (2, 0) gel phase peak was not instrumentally resolved, the correlation length of chains was greater than $426\,\text{Å}$. The (2, 0) Bragg peak width of a similar lipid, DPPC, was resolved and had a correlation length of $2900\,\text{Å}$ [50].

In contrast, the observed ripple phase WAXS peaks were instrumentally resolved (Fig. 3.53). The FWHM of the stronger peak was estimated to be $0.025\,\text{Å}^{-1}$, corresponding to a correlation length of $\sim240\,\text{Å}$, indicating that the correlation length in the ripple phase is shorter than that in the gel phase. This observation can be qualitatively understood by supposing that chains in the major and minor arms are not correlated, so that gel phase-like chains in the major arm are only correlated within the major arm, limiting the correlation length along the ripple direction to be less than the length of the major arm, $\sim100\,\text{Å}$. It is possible that chains are correlated over a much longer distance along the direction perpendicular to the ripple direction leading to a sharp reflection along q_y. In the case of our in-plane powder sample, we would observe the convolution of a broad width peak along q_x and a sharp peak along q_y. Such a convolution would result in a broad Bragg rod, qualitatively consistent with our nGIWAXS data. To quantitatively understand the observed peak widths would require rigorously modeling the finite size effect. This could lead to a prediction for the peak shape that is not Gaussian as assumed by the Scherrer equation [37].

Next, we compare our chain packing results to the result of the atomistic MD simulations by de Vries et al. [28]. In their simulations, while chains were straightened out (all-trans) like in the case of the gel phase, their chain tilt angles θ were modulated along the ripple direction. It was also clearly seen in their simulations that chains in the different leaflets were decoupled and tilted in opposite directions. Our tWAXS data are inconsistent with this picture and instead consistent with normal gel phase packing where chains in different leaflets constitute long coherently scattering entities.

3.11.2 Minor Arm

Some previous work has suggested that chains in the minor arm are disordered similarly to the L_α fluid phase [8, 27, 33, 43]. Figure 3.69 compares our electron density profiles in the minor arm to the DMPC fluid phase profile reported by Kučerka et al. [55]. Unlike the case for the major arm, the density profile in the minor arm is not quantitatively consistent with that of the fluid phase. However, as we discussed in Sect. 3.6, the electron density profile for the minor arm was less robust than for the major arm, so correcting the phase factors might lead to a profile more similar to the one shown in Fig. 3.69 (middle).

While the electron density profiles obtained through LAXS data analysis are not very supportive of the fluid phase like minor arm, our nGIWAXS data can be understood as scattering arising from the gel like major arm and the fluid like minor arm. As Figs. 3.53 and 3.54 show, the nGIWAXS pattern is consistent with a superposition of the two Bragg rods due to the major arm and a broad peak similar to the fluid phase WAXS pattern shown in Fig. 3.47. Table 3.19 shows that the width of the diffuse scattering peak in the fluid phase is approximately twice wider than the estimated width of the broad peak in the ripple phase. This difference can be understood by assuming that chains in the minor arm are less disordered than in the fluid phase, which would yield a narrower scattering feature. This assumption is reasonable because the ripple phase is a lower temperature phase than the fluid phase and the short length of the minor arm along the ripple direction might restrict the motion of disordered chains. This assumption would also explain why the estimated broad peak q-position is larger than that of the fluid phase WAXS peak (see Table 3.19); less chain disordering results in a smaller average chain-chain distance, leading to a larger q value of the associated scattering.

The existence of the broad peak in the ripple nGIWAXS is based on the assumption that the Bragg rod profiles in q_r are Lorentzians (see Figs. 3.53 and 3.54). Since the $L_{\beta I}$ (2, 0) Bragg rod profile in q_r was fitted very well with a Lorentzian (Fig. 3.48), this functional form is not unreasonable for the Bragg rods in the ripple phase. The thin rod model considered in Sect. 3.9.2 predicts a delta function in q_r, and the Scherrer particle broadening leads to a Gaussian profile [37]. Therefore, supporting a Lorentzian profile would require a more rigorous calculation of Bragg rod scattering including a finite size effect due to the major arm shape and/or disordering of chains discussed in Sect. 3.10.

An interesting chain packing shown in Fig. 3.70 was reported from an MD simulation [28] that introduced the possibility that the thinner minor arm might consist of interdigitated chains rather than disordered fluid chains. Figure 3.69 compares the electron density profiles in the minor arm with that of the dihexadecylphosphatidylcholine (DHPC) L_I interdigitated phase reported by [38]. Absence of the methyl trough can be seen in both L_I phase and the ripple minor arm, but the widths of the headgroups are much narrower in the DHPC L_I phase because the electron density of the backbone is closer to that of water because there is no carbonyl group. The widths of the "headgroups" in the minor arm profile are

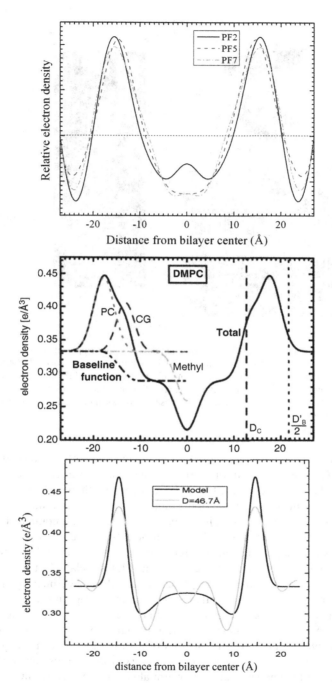

Fig. 3.69 Comparison of the electron density profiles of the minor arm in the ripple phase (*top*), the DMPC fluid phase (*middle*) reproduced from [55], and the DHPC interdigitated L_I phase (*bottom*) from [38]. The ripple minor arm profiles were calculated using the phase factors predicted by Fit2, 5, and 7 (*solid*, *dashed*, and *dash-dotted*, respectively). The *horizontal dashed line* in the *top panel* marks $\rho(z) = \rho_w$

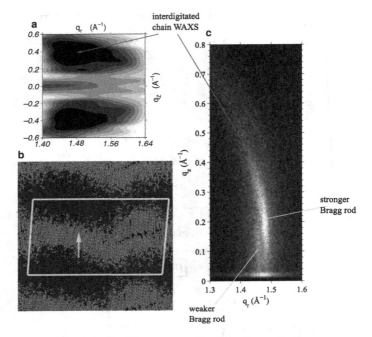

Fig. 3.70 Comparison of the WAXS pattern predicted by the ripple phase structure proposed in [28] (**a**, with *black* corresponding to highest intensity) and our measured WAXS (**c**, *white* corresponding to highest intensity). (**b**) Is a simulation snapshot from [28] that shows interdigitated chains in the minor arm. The *arrows* indicate the position of the maximal scattering observed in **a**, which is at $q_z \approx 0.4\,\text{Å}^{-1}$. The q_z values of the observed peaks in our data were 0.12 and $0.2\,\text{Å}^{-1}$, indicated by *arrows*

about 10 Å, comparable with those in the major arm profile shown in Fig. 3.67 and in the L_α phase shown in Fig. 3.69 (middle). Since each DHPC chain has two additional CH_2 groups compared to DMPC, the bilayer thickness would be smaller by 2×1.27 Å if the chains in the ripple minor arm were interdigitated.

While Fig. 3.69 suggests that interdigitated chains are not completely inconsistent with our results in terms of the overall density profile shape, our nGIWAXS pattern does not support the packing proposed in [28]. Figure 3.70 compares their calculated WAXS pattern from the atomistic MD simulations to our measured nGIWAXS pattern from the ripple phase. As noted in [28], interdigitated chains in the minor arm scatter coherently, giving rise to a Bragg rod centered at $q_z \approx 0.4\,\text{Å}^{-1}$. This off-equator Bragg rod would be due to chains being tilted by about 20° with respect to the stacking z direction though they are essentially parallel to the local bilayer normal. We saw no sign of a Bragg rod at such a large q_z value in our nGIWAXS data. Moreover, the two observed Bragg rods can be understood as arising from the chains in the major arm alone. Therefore, our nGIWAXS data do not support the structure proposed by de Vries et al. [28] although the electron density profiles in the minor arm shown in Fig. 3.69 are qualitatively consistent with an interdigitated structure. Resolving this conflict might require a more accurate electron density map with correct phase factors, and/or a rigorous calculation of a WAXS pattern from the minor arm, allowing a possibility to place an interdigitated chain lattice differently from the simulation as we did for the hexagonal lattice in the major arm using the variable ζ.

3.12 Conclusion

We have obtained a high resolution electron density map for the asymmetric DMPC ripple phase by analyzing the LAXS diffraction pattern collected with synchrotron X-rays. While the phase factors for some of the weak observed peaks remain ambiguous, the calculated electron density maps with various choices of the ambiguous phase factors clearly showed an out-of-plane structure similar to the gel phase in the major arm. They also showed that the minor arm electron density profiles do not resemble a typical fluid phase profile, but are more consistent with interdigitated chains. Measured ripple amplitudes, major and minor arm lengths and thicknesses did not depend considerably on the various phase factors predicted by different models and fits, and were therefore robust. Our analysis also confirms that the major arm is twice longer than the minor arm.

A further effort should be made to obtain the correct phase factors for the 12 uncertain orders with an approach similar to the pattern recognition method and/or development of a model that better captures the details of the ripple profile in the kink region. With the analysis detailed in this thesis, more data at various hydration levels could elucidate the inter- and intra-bilayer interactions in this phase. Measurements of the major arm thickness as a function of hydration would also indicate whether the chain tilt angle is dependent on the hydration as it is in the gel phase. If chains in the minor arm are indeed interdigitated, the minor arm thickness might be more or less constant.

Our unprecedented high resolution WAXS study on the ripple phase revealed for the first time clear separation of the previously observed broad peak [31]. The observed Bragg rods were indexed using our thin rod model developed for the ripple phase and additional information obtained from the LAXS analysis, showing that the chains in the major arm are tilted with respect to the bilayer normal (Fig. 3.68); this disagrees with the chain packing proposed in [11, 12]. By combining the structural parameters obtained from the LAXS and WAXS studies, we have suggested that chains in the major arm are slightly more disordered than all-trans chains. The Bragg rod indexing suggests that the observed Bragg rods arise from the major arm, and not from interdigitated chains in the minor arm suggested from a MD simulation. Our attempt in estimating the Bragg rod widths in the in-plane direction also showed a sign of broad scattering underneath the Bragg rods, which could be attributed to fluid like chains in the minor arm.

Future possible experiments include a high resolution transmission experiment, where both geometric broadening and energy dispersion are minimized. The expected resolution is the width of the X-ray beam, which is about three pixels. This experiment would double the tWAXS resolution achieved in this work. Another slightly different high resolution experiment is to use a silicon crystal analyzer downstream of the sample, which removes geometric broadening. The downside of this type of high resolution experiment is that only one point in q-space is probed by each exposure, so getting a full 2D map of wide angle scattering is time consuming. With high resolution nGIWAXS or tWAXS experiments close to full hydration, it might be possible to observe the $P_{\beta I}$ phase of the major arm. This would strongly support our analysis of the nGIWAXS data, which gives the $P_{\beta F}$ phase below full hydration. Combining high resolution LAXS and WAXS data at various hydration levels might be useful.

One could also calculate the WAXS pattern from the fluid like domains tilted with respect to the stacking z direction. In doing so, in-plane powder averaging should also be done, which would convolute the tilted ring patterns from tilted fluid domains in different directions. It would be interesting to see where this would locate the diffuse scattering maximum, on or off the equator.

Also highly speculative, but the ripple phase might be an interesting phase to study curvature sensing peptides. Curvature sensing peptides may accumulate at the kink regions. Then, the electron density profile can be calculated with the analysis detailed in this work. It would be very interesting if peptide-lipid interactions also significantly modify the wide angle pattern. With a known perturbation property of a peptide on lipids, it could shed light on the structure of the minor arm. For example, if indeed chains are fluid like in the minor arm, some peptides might accumulate in the minor arm because of the ease of insertion as compared to the gel-like major arm. Then, the ripple phase might be used to study biologically relevant problems.

In conclusion, the LAXS, nGIWAXS, and tWAXS analysis led to strong support for chain packing in the major arm similar to the gel phase with coupled leaflets, and clearly suggest different types of packing in the major and minor arms with conflicting results regarding the minor arm packing. Our study leads to more possible experiments and analyses, and should stimulate further research on the ripple phase, which continues to be mysterious and fascinating.

Bibliography

1. A. Tardieu, V. Luzzati, F. Reman, Structure and polymorphism of the hydrocarbon chains of lipids: a study of lecithin-water phases. J. Mol. Biol. **75**(4), 711–733 (1973)
2. R. Koynova, A. Koumanov, B. Tenchov, Metastable rippled gel phase in saturated phosphatidylcholines: calorimetric and densitometric characterization. Biochimica et Biophysica Acta (BBA) – Biomembranes **1285**(1), 101–108 (1996)
3. J. Katsaras, S. Tristram-Nagle, Y. Liu, R. Headrick, E. Fontes, P. Mason, J.F. Nagle, Clarification of the ripple phase of lecithin bilayers using fully hydrated, aligned samples. Phys. Rev. E **61**(5), 5668 (2000)

4. M.J. Janiak, D.M. Small, G.G. Shipley, Nature of the thermal pretransition of synthetic phospholipids: dimyristoyl- and dipalmitoyllecithin. Biochemistry 15(21), 4575–4580 (1976)
5. M.J. Janiak, D.M. Small, G.G. Shipley, Temperature and compositional dependence of the structure of hydrated dimyristoyl lecithin. J. Biol. Chem. 254(13), 6068–6078 (1979)
6. D.C. Wack, W.W. Webb, Synchrotron X-ray study of the modulated lamellar phase $P_{\beta'}$ in the lecithin-water system. Phys. Rev. A 40, 2712–2730 (1989)
7. H. Yao, S. Matuoka, B. Tenchov, I. Hatta, Metastable ripple phase of fully hydrated dipalmitoylphosphatidylcholine as studied by small angle X-ray scattering. Biophys. J. 59(1), 252–255 (1991)
8. W.J. Sun, S. Tristram-Nagle, R.M. Suter, J.F. Nagle, Structure of the ripple phase in lecithin bilayers. Proc. Natl. Acad. Sci. 93(14), 7008–7012 (1996)
9. B.A. Cunningham, A.-D. Brown, D.H. Wolfe, W.P. Williams, A. Brain, Ripple phase formation in phosphatidylcholine: effect of acyl chain relative length, position, and unsaturation. Phys. Rev. E 58(3), 3662 (1998)
10. K. Sengupta, V. Raghunathan, J. Katsaras, Structure of the ripple phase in chiral and racemic dimyristoylphosphatidylcholine multibilayers. Phys. Rev. E 59(2), 2455 (1999)
11. K. Sengupta, V. Raghunathan, J. Katsaras, Novel structural features of the ripple phase of phospholipids. Europhys. Lett. 49(6), 722 (2000)
12. K. Sengupta, V.A. Raghunathan, J. Katsaras, Structure of the ripple phase of phospholipid multibilayers. Phys. Rev. E 68, 031710 (2003)
13. K. Mortensen, W. Pfeiffer, E. Sackmann, W. Knoll, Structural properties of a phosphatidylcholine-cholesterol system as studied by small-angle neutron scattering: ripple structure and phase diagram. Biochimica et Biophysica Acta (BBA)-Biomembranes 945(2), 221–245 (1988)
14. J.P. Bradshaw, M.S. Edenborough, P.J. Sizer, A. Watts, Observation of rippled dioleoylphosphatidylcholine bilayers by neutron diffraction. Biochimica et Biophysica Acta (BBA)-Biomembranes 987(1), 111–114 (1989)
15. P.C. Mason, B.D. Gaulin, R.M. Epand, G.D. Wignall, J.S. Lin, Small angle neutron scattering and calorimetric studies of large unilamellar vesicles of the phospholipid dipalmitoylphosphatidylcholine. Phys. Rev. E 59(3), 3361 (1999)
16. J. Woodward IV, J. Zasadzinski, Amplitude, wave form, and temperature dependence of bilayer ripples in the $P_{\beta'}$ phase. Phys. Rev. E 53(4), R3044 (1996)
17. B.R. Copeland, H.M. McConnell, The rippled structure in bilayer membranes of phosphatidylcholine and binary mixtures of phosphatidylcholine and cholesterol. Biochimica et Biophysica Acta (BBA) – Biomembranes 599(1), 95–109 (1980)
18. D. Ruppel, E. Sackmann, On defects in different phases of two-dimensional lipid bilayers. Journal de Physique 44(9), 1025–1034 (1983)
19. J. Zasadzinski, J. Schneir, J. Gurley, V. Elings, P. Hansma, Scanning tunneling microscopy of freeze-fracture replicas of biomembranes. Science 239(4843), 1013–1015 (1988)
20. R.A. Parente, B.R. Lentz, Phase behavior of large unilamellar vesicles composed of synthetic phospholipids. Biochemistry 23(11), 2353–2362 (1984)
21. L. Li, J.-X. Cheng, Coexisting stripe-and patch-shaped domains in giant unilamellar vesicles. Biochemistry 45(39), 11819–11826 (2006)
22. J. Nagle, Theory of lipid monolayer and bilayer phase transitions: effect of headgroup interactions. J. Membr. Biol. 27(1), 233–250 (1976)
23. T.J. McIntosh, Differences in hydrocarbon chain tilt between hydrated phosphatidylethanolamine and phosphatidylcholine bilayers. A molecular packing model. Biophys. J. 29(2), 237–245 (1980)
24. J.F. Nagle, S. Tristram-Nagle, Structure of lipid bilayers. Biochimica et Biophysica Acta (BBA) – Reviews on Biomembranes 1469(3), 159–195 (2000)
25. M.P. Hentschel, F. Rustichelli, Structure of the ripple phase $P_{\beta'}$ in hydrated phosphatidylcholine multimembranes. Phys. Rev. Lett. 66, 903–906 (1991)
26. R. Wittebort, C. Schmidt, R. Griffin, Solid-state carbon-13 nuclear magnetic resonance of the lecithin gel to liquid-crystalline phase transition. Biochemistry 20(14), 4223–4228 (1981)

27. M.B. Schneider, W.K. Chan, W.W. Webb, Fast diffusion along defects and corrugations in phospholipid $P_{\beta'}$ liquid crystals. Biophys. J. **43**(2), 157–165 (1983)
28. A.H. de Vries, S. Yefimov, A.E. Mark, S.J. Marrink, Molecular structure of the lecithin ripple phase. Proc. Natl. Acad. Sci. **102**(15), 5392–5396 (2005)
29. O. Lenz, F. Schmid, Structure of symmetric and asymmetric "ripple" phases in lipid bilayers. Phys. Rev. Lett. **98**, 058104 (2007)
30. C.M. Chen, T.C. Lubensky, F.C. MacKintosh, Phase transitions and modulated phases in lipid bilayers. Phys. Rev. E **51**(1), 504 (1995)
31. J. Katsaras, V.A. Raghunathan, Molecular chirality and the "ripple" phase of phosphatidyl-choline multibilayers. Phys. Rev. Lett. **74**, 2022–2025 (1995)
32. M.A. Kamal, A. Pal, V.A. Raghunathan, M. Rao, Theory of the asymmetric ripple phase in achiral lipid membranes. Europhys. Lett. **95**(4), 48004 (2011)
33. K.A. Riske, R.P. Barroso, C.C. Vequi-Suplicy, R. Germano, V.B. Henriques, M.T. Lamy, Lipid bilayer pre-transition as the beginning of the melting process. Biochimica et Biophysica Acta (BBA) – Biomembranes **1788**(5), 954–963 (2009)
34. S.A. Tristram-Nagle, Preparation of oriented, fully hydrated lipid samples for structure determination using X-ray scattering. Methods Mol. Biol. **400**, 63–75 (2007)
35. http://henke.lbl.gov/optical_constants
36. Y. Liu, New method to obtain strcuture of biomembranes using diffuse X-ray scattering: application to fluid phase DOPC lipid bilayers. PhD thesis, Carnegie Mellon University (2003)
37. B.E. Warren, *X-ray Diffraction* (Courier Dover Publications, New York, 1969)
38. S. Guler, D.D. Ghosh, J. Pan, J.C. Mathai, M.L. Zeidel, J.F. Nagle, S. Tristram-Nagle, Effects of ether vs. ester linkage on lipid bilayer structure and water permeability. Chem. Phys. Lipids **160**(1), 33–44 (2009)
39. http://cars9.uchicago.edu/software/python/lmfit/
40. M.C. Wiener, R.M. Suter, J.F. Nagle, Structure of the fully hydrated gel phase of dipalmi-toylphosphatidylcholine. Biophys. J. **55**(2), 315–325 (1989)
41. C. Worthington, G. King, T. McIntosh, Direct structure determination of multilayered membrane-type systems which contain fluid layers. Biophys. J. **13**(5), 480–494 (1973)
42. S. Tristram-Nagle, Y. Liu, J. Legleiter, J.F. Nagle, Structure of gel phase DMPC determined by X-ray diffraction. Biophys. J. **83**(6), 3324–3335 (2002)
43. G. Pabst, H. Amenitsch, D.P. Kharakoz, P. Laggner, M. Rappolt, Structure and fluctuations of phosphatidylcholines in the vicinity of the main phase transition. Phys. Rev. E **70**(2), 021908 (2004)
44. T.T. Mills, G.E. Toombes, S. Tristram-Nagle, D.-M. Smilgies, G.W. Feigenson, J.F. Nagle, Order parameters and areas in fluid-phase oriented lipid membranes using wide angle X-ray scattering. Biophys. J. **95**(2), 669–681 (2008)
45. G.S. Smith, E.B. Sirota, C.R. Safinya, N.A. Clark, Structure of the $L_{\beta'}$ phases in a hydrated phosphatidylcholine multimembrane. Phys. Rev. Lett. **60**, 813–816 (1988)
46. G.H. Vineyard, Grazing-incidence diffraction and the distorted-wave approximation for the study of surfaces. Phys. Rev. B **26**(8), 4146 (1982)
47. C.E. Miller, J. Majewski, E.B. Watkins, D.J. Mulder, T. Gog, T.L. Kuhl, Probing the local order of single phospholipid membranes using grazing incidence X-ray diffraction. Phys. Rev. Lett. **100**(5), 058103 (2008)
48. K. Akabori, J.F. Nagle, Comparing lipid membranes in different environments. ACS Nano **8**(4), 3123–3127 (2014)
49. S. Tristram-Nagle, R. Zhang, R.M. Suter, C.R. Worthington, W.J. Sun, J.F. Nagle, Measure-ment of chain tilt angle in fully hydrated bilayers of gel phase lecithins. Biophys. J. **64**(4), 1097–1109 (1993)
50. W.J. Sun, R.M. Suter, M.A. Knewtson, C.R. Worthington, S. Tristram-Nagle, R. Zhang, J.F. Nagle, Order and disorder in fully hydrated unoriented bilayers of gel-phase dipalmi-toylphosphatidylcholine. Phys. Rev. E **49**, 4665–4676 (1994)
51. N. Chu, N. Kučerka, Y. Liu, S. Tristram-Nagle, J.F. Nagle, Anomalous swelling of lipid bilayer stacks is caused by softening of the bending modulus. Phys. Rev. E **71**, 041904 (2005)

52. H. Träuble, D.H. Haynes, The volume change in lipid bilayer lamellae at the crystalline-liquid crystalline phase transition. Chem. Phys. Lipids **7**(4), 324–335 (1971)
53. J.F. Nagle, Theory of the main lipid bilayer phase-transition. Annu. Rev. Phys. Chem. **31**, 157–195 (1980)
54. E.B. Watkins, C.E. Miller, W.-P. Liao, T.L. Kuhl, Equilibrium or quenched: fundamental differences between lipid monolayers, supported bilayers, and membranes. ACS Nano **8**(4), 3181–3191 (2014)
55. N. Kučerka, Y. Liu, N. Chu, H.I. Petrache, S. Tristram-Nagle, J.F. Nagle, Structure of fully hydrated fluid phase DMPC and DLPC lipid bilayers using X-ray scattering from oriented multilamellar arrays and from unilamellar vesicles. Biophys. J. **88**(4), 2626–2637 (2005)

Appendices

Appendix A

Abstract In this appendix, we develop an analytical framework for dealing with mosaic spread and discuss experimental techniques to measure the degree of mosaic spread.

A.1 Mosaic Spread for NFIT Analysis

First we calculate how mosaic spread affects the structure factor S(q). Next we discuss two experimental methods. Third, we discuss the updated NFIT program. Fourth, we show the results.

A.1.1 Mosaic Spread: Calculation

In this section, an analytical framework for dealing with mosaic spread is developed. A sample of oriented stacks of bilayers consists of many small domains, within which layers are registered in an array. An ideal domain is a domain where the layers are parallel to the substrate, whose surface is in the sample xy-plane, so the orientation $\mathbf{n}$ of an ideal domain is perpendicular to the substrate as shown in Fig. A.1. In general, the orientation $\mathbf{n}'$ of a domain is tilted from that of an ideal domain by some angle α. Then, we consider a mosaic spread distribution function, $P(\alpha)$, representing a probability of finding a domain with a tilt α. We assume that the sample is symmetric about the substrate normal, so that the distribution $P(\alpha)$ does not depend on the azimuthal angle, β. The normalization condition on $P(\alpha)$ is

$$1 = \int_0^{2\pi} d\beta \int_0^{\frac{\pi}{2}} d\alpha \, \sin \alpha P(\alpha). \tag{A.1}$$

The object of this section is to derive the X-ray scattering structure factor including the distribution function $P(\alpha)$.

First, let us consider a two dimensional example. Our sample consists of two identical domains except a tilt α shown in Fig. A.2. Then, the sample structure factor

© Springer International Publishing Switzerland 2015

K. Akabori, *Structure Determination of HIV-1 Tat/Fluid Phase Membranes and DMPC Ripple Phase Using X-Ray Scattering*, Springer Theses, DOI 10.1007/978-3-319-22210-3

157

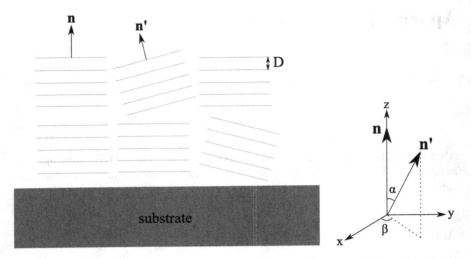

Fig. A.1 Two dimensional view of mosaic spread (*left*) and notations used in this section (*right*). The stacking direction of an ideal domain is **n** and that of a tilted domain **n'**. The deviation of **n'** from **n** denoted as α quantifies the degree of misorientation of a domain. The x, y, and z-axes are the sample coordinates

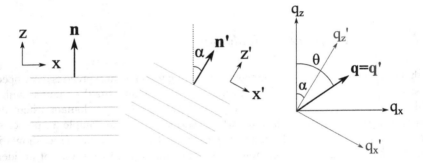

Fig. A.2 Example of a two dimensional sample consisting of an ideal and tilted domains. $\mathbf{q} = (q_x, q_z)$ is the sample q-space and $\mathbf{q'} = (q'_x, q'_z)$ is the domain q-space. The two q-spaces are related by a rotation of α about the y-axis, which is into the page

$S^{\text{sam}}(\mathbf{q})$ is a superposition of the structure factor $S(\mathbf{q})$ of the ideal domain and $S(\mathbf{q'})$ of the tilted domain,

$$S^{\text{sam}}(\mathbf{q}) = S(q_x, q_z) + S(q'_x, q'_z). \tag{A.2}$$

To express $S(q'_x, q'_z)$ in terms of the sample q-space (q_x, q_z), we write q'_x and q'_z in terms of q_x, q_z, and α,

$$q'_x = \mathbf{q} \cdot \hat{\mathbf{x}}' = q \cos\left(\frac{\pi}{2} - \theta + \alpha\right)$$

$$q'_z = \mathbf{q} \cdot \hat{\mathbf{z}}' = q \sin\left(\frac{\pi}{2} - \theta + \alpha\right)$$

$$q_x = q \cos(\pi/2 - \theta)$$

$$q_z = q \sin(\pi/2 - \theta) \qquad (A.3)$$

where $q = |\mathbf{q}|$. Equations (A.2) and (A.3) give the structure factor of a sample consisting of the two domains. With a continuous distribution of $\mathbf{n}'$, we integrate over the angle α with each structure factor modulated by the distribution function $P(\alpha)$,

$$S_M(\mathbf{q}) = S_M(q, \theta) = \int_{-\frac{\pi}{2}}^{\frac{\pi}{2}} d\alpha \, S(q'_x, q'_z) P(\alpha), \qquad (A.4)$$

Variables q and θ are used in the above equation to make a connection with the three dimensional case, where the spherical coordinates are convenient, which we discuss now.

For a three dimensional sample, the basic idea is the same as the two dimensional case. In the three dimensional case, we also rotate the vector $\mathbf{n}'$ about the z-axis by an angle β after the rotation about the y-axis by an angle α, so all we need to do is to apply appropriate rotation matrices to the sample xyz-axes which define the domain coordinates $x'y'z'$.

The rotation matrix for rotating a vector about the y-axis is given by

$$R_y = \begin{pmatrix} \cos\alpha & 0 & \sin\alpha \\ 0 & 1 & 0 \\ -\sin\alpha & 0 & \cos\alpha \end{pmatrix} \qquad (A.5)$$

and for rotating about the z-axis

$$R_z = \begin{pmatrix} \cos\beta & -\sin\beta & 0 \\ \sin\beta & \cos\beta & 0 \\ 0 & 0 & 1 \end{pmatrix}. \qquad (A.6)$$

Then, what we want is

$$\hat{\mathbf{x}}' = R_z R_y \begin{pmatrix} 1 \\ 0 \\ 0 \end{pmatrix} = \begin{pmatrix} \cos\alpha \cos\beta \\ \cos\alpha \sin\beta \\ -\sin\alpha \end{pmatrix} \qquad (A.7)$$

$$\hat{\mathbf{y}}' = R_z R_y \begin{pmatrix} 0 \\ 1 \\ 0 \end{pmatrix} = \begin{pmatrix} -\sin\beta \\ \cos\beta \\ 0 \end{pmatrix} \qquad (A.8)$$

$$\hat{\mathbf{z}}' = R_z R_y \begin{pmatrix} 0 \\ 0 \\ 1 \end{pmatrix} = \begin{pmatrix} \sin\alpha\cos\beta \\ \sin\alpha\sin\beta \\ \cos\alpha \end{pmatrix}. \tag{A.9}$$

The domain q-space, (q'_x, q'_y, q'_z), in terms of the sample q-space (q_x, q_y, q_z) is given by

$$q'_x = \mathbf{q} \cdot \hat{\mathbf{x}}' = q_x \cos\alpha\cos\beta + q_y \cos\alpha\sin\beta - q_z \sin\alpha, \tag{A.10}$$

$$q'_y = \mathbf{q} \cdot \hat{\mathbf{y}}' = -q_x \sin\beta + q_y \cos\beta, \tag{A.11}$$

$$q'_z = \mathbf{q} \cdot \hat{\mathbf{z}}' = q_x \sin\alpha\cos\beta + q_y \sin\alpha\sin\beta + q_z \cos\alpha. \tag{A.12}$$

The transformation expressed in the spherical coordinates is

$$\cos\theta' = \frac{q'_z}{q} = \sin\theta\sin\alpha\cos(\phi - \beta) + \cos\theta\cos\alpha, \tag{A.13}$$

$$\tan\phi' = \frac{q'_y}{q'_x} = \frac{\sin\theta\sin(\phi - \beta)}{\sin\theta\cos\alpha\cos(\phi - \beta) - \cos\theta\sin\alpha}. \tag{A.14}$$

Summing over all the domains, we get for the mosaic spread modified structure factor

$$S_M(q, \theta, \phi) = \int_0^{2\pi} d\beta \int_0^{\frac{\pi}{2}} d\alpha \, S(q, \theta', \phi') P(\alpha) \tag{A.15}$$

with Eqs. (A.13) and (A.14).

To test these equations, let us apply them to the simple case of a stack of rigid layers with their normals parallel to the z-axis in spherical coordinates. The structure factor is then

$$S(q, \theta, \phi) = \frac{\delta(q - \frac{2\pi h}{D})}{q^2} \delta(\cos\theta - 1)\delta(\phi) \tag{A.16}$$

where $\delta(x)$ is the Dirac delta function. From Eq. (A.14), $\delta(\phi')$ is equivalent to $\delta(\beta - \phi)$. Setting $\beta = \phi$ in Eq. (A.13) gives $\cos\theta' = \cos(\alpha - \theta)$. Then, the mosaic spread modified structure factor $S_M(\mathbf{q})$ is

$$S_M(q, \theta, \phi) = \int d\alpha \int d\beta \, \frac{\delta(q - \frac{2\pi h}{D})}{q^2} \delta(\cos\theta' - 1)\delta(\beta - \phi) P(\alpha)$$

$$= \frac{\delta(q - \frac{2\pi h}{D})}{q^2} \int d\alpha \, \delta(\cos[\alpha - \theta] - 1) P(\alpha)$$

$$= \frac{\delta(q - \frac{2\pi h}{D})}{q^2} P(\theta). \tag{A.17}$$

Equation (A.17) describes hemispherical shells with radii of $2\pi h/D$ in the sample q-space. As will be described in the next section, a 2D detector records cross sections of these shells, which give rise to mosaic arcs along $q = 2\pi h/D$.

The structure factor of thermally fluctuating layers is not simple delta functions and gives rise to diffuse scattering. Analysis of the diffuse scattering from a sample with mosaic spread requires Eq. (A.15).

A.1.2 Mosaic Spread: Near Equivalence of Two Methods

In this section, we discuss experimental procedures to probe appropriate q-space to measure the mosaic spread distribution, $P(\alpha)$. In our setup, the angle of incidence between the beam and substrate, denoted by ω, can be varied. A conventional method to measure $P(\alpha)$ is a rocking scan, where one measures the integrated intensity of a given Bragg peak as a function of ω with a fixed detector position. Another method that takes an advantage of an area detector [1] measures the intensity as a function of χ on a two dimensional detector (see Fig. A.3). This method has been used to quantify complete pole figures for thin films with fiber texture (isotropic in-plane orientation) [2]. First, we want to compare the two methods mentioned above and determine their relationship.

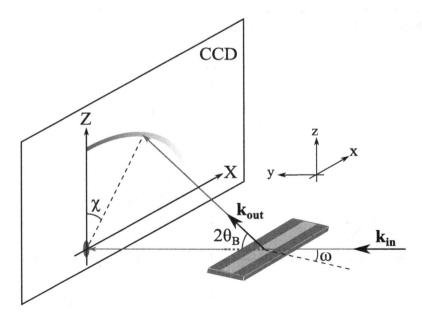

Fig. A.3 Notations used in this section. The arc originating from the Z-axis is the mosaic arc due to the mosaic spread distribution

Equation (3.6) expressed in terms of the coordinates defined in Fig. A.3 is

$$q_x = q \cos \theta \sin \chi$$
$$q_y = q \left(- \sin \theta \cos \omega + \cos \theta \cos \chi \sin \omega \right)$$
$$q_z = q \left(\sin \theta \sin \omega + \cos \theta \cos \chi \cos \omega \right). \qquad (A.18)$$

For a rocking scan focused on a particular order, $\chi = 0$ and $\theta = \theta_B$ while ω is varied about θ_B, where θ_B is the Bragg angle. Then,

$$q_x = 0$$
$$q_y = q_B \sin(\omega - \theta_B)$$
$$q_z = q_B \cos(\omega - \theta_B), \qquad (A.19)$$

which shows that this scan traces a part of the circular path in the $q_x = 0$ plane as shown in Fig. A.4. As Fig. A.4 shows, however, the rocking scan only probes a small fraction of the entire distribution, limited by $2\theta_B$. As discussed in Sect. 3.3.2, beyond $\omega = 2\theta_B$, the substrate blocks scattering. On the other hand, the ring analysis takes advantage of a two dimensional detector and can probe a substantially wider range of the distribution in principle: approximately $\pm 45°$ at $\omega = \theta_B$. This method is now described.

In the ring method, we set $\omega = \theta_B$ and scan on the detector along $\theta = \theta_B$ as a function of χ. Then, Eq. (A.18) becomes

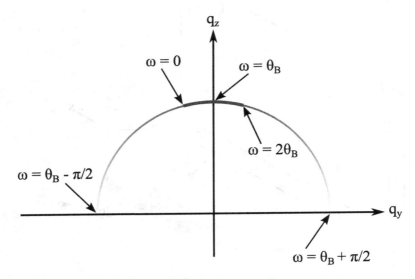

Fig. A.4 Rocking scan trace in q-space

$$q_x = q \cos \theta_B \sin \chi$$
$$q_y = q \sin \theta_B \cos \theta_B (\cos \chi - 1)$$
$$q_z = q(\sin^2 \theta_B + \cos^2 \theta_B \cos \chi), \qquad (A.20)$$

where $q = 4\pi \sin \theta_B / \lambda$. For small θ_B, Eq. (A.20) reduces to

$$q_x \approx q \sin \chi$$
$$q_y \approx 0$$
$$q_z \approx q \cos \chi. \qquad (A.21)$$

For a sharp Bragg peak, this ring method gives the same mosaic intensity $I(\chi, \theta_B)$ in Eq. (A.21) as the rocking method mosaic intensity $I(\omega - \theta_B)$ in Eq. (A.19) because the mosaic distribution $P(\alpha)$ is in-plane isotropic. Differences occur when diffuse scattering is added. The diffuse scattering intensity is much broader and weaker than the Bragg peaks. In the ring method, it can be estimated as the average from two rings offset on either side from θ_B and subtracted from the θ_B ring.

A.1.3 NFIT

The original NFIT program was written by Dr. Yufeng Liu and described in his thesis. It was used in the Nagle lab, with small updates for data handling, from 2003 until recently. A newer version has been implemented by Michael Jablin that calculates the theoretical structure factor using cylindrical domains appropriate for in-plane correlations [3] rather than rectangular domains appropriate for coherence domains. All these versions approximated the effect of mosaic spread roughly by averaging only in the q_r direction at fixed q_z which means that mosaic rings are approximated as mosaic lines or spikes. The subsequent development described here and not yet adopted by the Nagle lab calculates the structure factor $S(q_r, q_z)$ with rotational symmetry about the z-axis, which eliminates the ϕ' dependence in Eq. (A.15). The program interpolates $S(q_r, q_z)$ in terms of the spherical coordinates q and θ with $\phi = 0$ to perform the double integration in Eq. (A.15). After the mosaic spread integration, the program performs the q_y integration described in Sect. 2.2.5. For this integration, the calculated S_M is interpolated in terms of q_x, q_y, and q_z.

Note: if the structure factor defined in the Cartesian coordinates is desired (for a case of square domains instead of circular ones), Eqs. (A.10), (A.11), and (A.12) can be used instead of Eqs. (A.13) and (A.14).

While it is an improvement, the new program also is an approximation because it does not include the unknown form factor $|F(q_z)|$. The mosaic spread integration mixes up intensity at different q_z values, so the separation of $|F(q_z)|$ from $S(\mathbf{q})$ is in principle impossible. One way to deal with this issue would be to combine the

SDP program, which determines $|F(q_z)|$, with the NFIT program, but that will end up with too many non-linear parameters. Another possibility is to limit the fitting range to regions close to the meridian. For a small range of integration, it is not unreasonable to assume that the form factor is approximately constant as can be seen from Eq. (A.12) with small q_x, q_y, and α. Therefore, the analysis developed in this appendix ignores the form factor.

A.2 Derivation of the Contour Part of the Form Factor

In this section, we derive F_C. The ripple profile, $u(x)$ is given by

$$
u(x) = \begin{cases} -\frac{A}{\lambda_r - x_0}\left(x + \frac{\lambda_r}{2}\right) & \text{for } -\frac{\lambda_r}{2} \leq x < -\frac{x_0}{2} \\ \frac{A}{x_0}x & \text{for } -\frac{x_0}{2} \leq x \leq \frac{x_0}{2} \\ -\frac{A}{\lambda_r - x_0}\left(x - \frac{\lambda_r}{2}\right) & \text{for } \frac{x_0}{2} < x \leq \frac{\lambda_r}{2} \end{cases}
\tag{A.22}
$$

The contour part of the form factor is the Fourier transform of the contour function, $C(x,z)$,

$$
F_C(\mathbf{q}) = \frac{1}{\lambda_r}\int_{-\frac{\lambda_r}{2}}^{\frac{\lambda_r}{2}} dx \int_{-\frac{D}{2}}^{\frac{D}{2}} dz\, C(x,z) e^{iq_z z} e^{iq_x x}
$$

As discussed in section X, the modulated models allow the electron density to modulate along the ripple direction, x. This means

$$
C(x,z) = \begin{cases} f_1 \delta[z - u(x)] & \text{for } -\frac{\lambda_r}{2} \leq x < -\frac{x_0}{2} \\ \delta[z - u(x)] & \text{for } -\frac{x_0}{2} < x < \frac{x_0}{2} \\ f_1 \delta[z - u(x)] & \text{for } \frac{x_0}{2} \leq x < \frac{\lambda_r}{2} \end{cases}
$$
$$
+ f_2\, \delta\!\left(x + \frac{x_0}{2}\right)\delta\!\left(z + \frac{A}{2}\right) + f_2\, \delta\!\left(x - \frac{x_0}{2}\right)\delta\!\left(z - \frac{A}{2}\right).
\tag{A.23}
$$

The contribution from the minor arm is

$$
\frac{1}{\lambda_r}\int_{-\frac{\lambda_r}{2}}^{-\frac{x_0}{2}} dx\, e^{iq_x x} e^{iq_z u(x)} + \int_{\frac{x_0}{2}}^{\frac{\lambda_r}{2}} dx\, e^{iq_x x} e^{iq_z u(x)}
$$
$$
= \frac{1}{\lambda_r}\int_{\frac{x_0}{2}}^{\frac{\lambda_r}{2}} dx\, e^{-i\left[q_x x - q_z \frac{A}{\lambda_r - x_0}\left(x - \frac{\lambda_r}{2}\right)\right]} + \int_{\frac{x_0}{2}}^{\frac{\lambda_r}{2}} dx\, e^{i\left[q_x x - q_z \frac{A}{\lambda_r - x_0}\left(x - \frac{\lambda_r}{2}\right)\right]}
$$
$$
= \frac{2}{\lambda_r}\int_{\frac{x_0}{2}}^{\frac{\lambda_r}{2}} \cos\left[\left(q_x - q_z \frac{A}{\lambda_r - x_0}\right)x + q_z \frac{A}{\lambda_r - x_0}\frac{\lambda_r}{2}\right]
\tag{A.24}
$$

Using a trigonometric identity,

$$\sin u - \sin v = 2 \cos[(u + v)/2] \sin[(u - v)/2],$$

and defining

$$\omega(\mathbf{q}) = \frac{1}{2} \left(q_x x_0 + q_z A \right), \tag{A.25}$$

we further simplify Eq. (A.24),

$$= \frac{2}{\lambda_r} \frac{\lambda_r - x_0}{\frac{1}{2} q_x \lambda_r - \omega} \cos\left[\frac{1}{2} \left(\frac{1}{2} q_x \lambda_r + \omega \right) \right] \sin\left[\frac{1}{2} \left(\frac{1}{2} q_x \lambda_r - \omega \right) \right]$$

$$= \frac{1}{\lambda_r} \frac{\lambda_r - x_0}{\frac{1}{2} q_x \lambda_r - \omega} \cos\left[\frac{1}{2} \left(\frac{1}{2} q_x \lambda_r + \omega \right) \right] \frac{\sin\left(\frac{1}{2} q_x \lambda_r - \omega \right)}{\cos\left[\frac{1}{2} \left(\frac{1}{2} q_x \lambda_r - \omega \right) \right]}$$

$$= \frac{\lambda_r - x_0}{\lambda_r} \frac{\cos\left[\frac{1}{2} \left(\frac{1}{2} q_x \lambda_r + \omega \right) \right]}{\cos\left[\frac{1}{2} \left(\frac{1}{2} q_x \lambda_r - \omega \right) \right]} \frac{\sin\left(\frac{1}{2} q_x \lambda_r - \omega \right)}{\frac{1}{2} q_x \lambda_r - \omega}. \tag{A.26}$$

Similarly, we calculate the contribution from the major arm,

$$\frac{1}{\lambda_r} \int_{-\frac{x_0}{2}}^{\frac{x_0}{2}} dx\, e^{i\left(\frac{q_z A}{x_0} + q_x \right) x} = \frac{2}{\lambda_r} \int_0^{\frac{x_0}{2}} dx \cos\left(\frac{q_z A}{x_0} + q_x \right) x$$

$$= \frac{x_0}{\lambda_r} \frac{\sin \omega}{\omega} \tag{A.27}$$

The contribution from the kink region is

$$\frac{1}{\lambda_r} \iint dx\, dz \left[\delta\left(x + \frac{x_0}{2}\right) \delta\left(z + \frac{A}{2}\right) + \delta\left(x - \frac{x_0}{2}\right) \delta\left(z - \frac{A}{2}\right) \right] e^{iq_x x} e^{iq_z z}$$

$$= \frac{2}{\lambda_r} \cos \omega. \tag{A.28}$$

Therefore,

$$F_C(\mathbf{q}) = \frac{x_0}{\lambda_r} \frac{\sin \omega}{\omega} + f_1 \frac{\lambda_r - x_0}{\lambda_r} \frac{\cos\left[\frac{1}{2} \left(\frac{1}{2} q_x \lambda_r + \omega \right) \right]}{\cos\left[\frac{1}{2} \left(\frac{1}{2} q_x \lambda_r - \omega \right) \right]} \frac{\sin\left(\frac{1}{2} q_x \lambda_r - \omega \right)}{\frac{1}{2} q_x \lambda_r - \omega}$$

$$+ \frac{2 f_2}{\lambda_r} \cos \omega \tag{A.29}$$

To allow different transbilayer models for the major and minor arms, we can write the form factor as

$$F(\mathbf{q}) = F_C^M(\mathbf{q}) F_T^M(\mathbf{q}) + f_1 F_C^m(\mathbf{q}) F_T^m(\mathbf{q}) + f_2 F_C^k(\mathbf{q}) F_T^k(\mathbf{q}) \tag{A.30}$$

such that

$$F_C^M = \frac{x_0}{\lambda_r} \frac{\sin \omega}{\omega} \tag{A.31}$$

$$F_C^m = \frac{\lambda_r - x_0}{\lambda_r} \frac{\cos\left[\frac{1}{2}\left(\frac{1}{2}q_x\lambda_r + \omega\right)\right]}{\cos\left[\frac{1}{2}\left(\frac{1}{2}q_x\lambda_r - \omega\right)\right]} \frac{\sin\left(\frac{1}{2}q_x\lambda_r - \omega\right)}{\frac{1}{2}q_x\lambda_r - \omega} \tag{A.32}$$

$$F_C^k = \frac{2}{\lambda_r} \cos \omega. \tag{A.33}$$

In this thesis, we employed the same model for F_T^M, F_T^m, and F_T^k, but one could also implement a gel phase model for F_T^M and interdigitated and fluid phase models for F_T^m.

A.3 Derivation of the Transbilayer Part of the Form Factor in the 2G Hybrid Model

In this section, we derive the transbilayer part of the form factor calculated from the 2G hybrid model discussed in Sect. 3.5. Defining $z' = -x\sin\psi + z\cos\psi$, the Fourier transform of a Gaussian function along the line tilted from z-axis by ψ is

$$\iint dz\, dx\, \rho_{Hi} \exp\left\{-\frac{(z' - Z_{Hi})^2}{2\sigma_{Hi}^2}\right\} \delta(x\cos\psi + z\sin\psi) e^{iq_x x} e^{iq_z z}$$

$$= \frac{1}{\cos\psi} \int_{-\frac{D}{2}}^{\frac{D}{2}} dz\, \rho_{Hi} \exp\left\{-\frac{(z - Z_{Hi}\cos\psi)^2}{2\sigma_{Hi}^2 \cos^2\psi} + i(q_z - q_x\tan\psi)z\right\}$$

$$\approx \rho_{Hi}\sqrt{2\pi}\sigma_{Hi} \exp\left\{i\alpha Z_{Hi} - \frac{1}{2}\alpha^2\sigma_{Hi}^2\right\} \tag{A.34}$$

with $\alpha = q_z\cos\psi - q_x\sin\psi$. Using Eq. (A.34) and adding the other side of the bilayer and the terminal methyl term, we get

$$F_G = \sqrt{2\pi}\left[-\rho_M\sigma_M \exp\left\{-\frac{1}{2}\alpha^2\sigma_M^2\right\}\right.$$

$$\left. + \sum_{i=1}^{1\text{ or }2} 2\rho_{Hi}\sigma_{Hi}\cos(\alpha Z_{Hi})\exp\left\{-\frac{1}{2}\alpha^2\sigma_{Hi}^2\right\}\right]. \tag{A.35}$$

The strip part of the model in the minus fluid convention is

$$\rho_S(z) = \begin{cases} -\Delta\rho & \text{for } 0 \leq z < Z_{CH_2}\cos\psi, \\ 0 & \text{for } Z_W\cos\psi \leq z \leq D/2, \end{cases} \tag{A.36}$$

where $\Delta\rho = \rho_W - \rho_{CH_2}$. Then, the corresponding Fourier transform is

$$F_S = \iint dz\,dx\,e^{iq_xx}e^{iq_zz}\rho_S(z)\delta(x\cos\psi + z\sin\psi)$$

$$= \frac{2}{\cos\psi}\int_0^{Z_{CH_2}\cos\psi} dz\cos\left(\frac{\alpha}{\cos\psi}z\right)(-\Delta\rho)$$

$$= -2\Delta\rho\frac{\sin(\alpha Z_{CH_2})}{\alpha}. \tag{A.37}$$

The bridging part of the model in the minus fluid convention is

$$\rho_B(x,z) = \frac{\Delta\rho}{2}\cos\left[\frac{-\pi}{\Delta Z_H}(z' - Z_W)\right] - \frac{\Delta\rho}{2} \tag{A.38}$$

for $Z_{CH_2}\cos\psi < z < Z_W\cos\psi$, and 0 otherwise. Here, $\Delta Z_H = Z_W - Z_{CH_2}$. Then, for the strip part of the form factor, we have

$$F_B = \iint dz\,dx\,e^{iq_xx}e^{iq_zz}\delta(x\cos\psi + z\sin\psi)\rho_B(x,z)$$

$$= \frac{\Delta\rho}{\cos\psi}\int_{Z_{CH_2}\cos\psi}^{Z_W\cos\psi} dz\cos\left(\alpha\frac{z}{\cos\psi}\right)\left\{\cos\left[-\frac{\pi}{\Delta Z_H}\left(\frac{z}{\cos\psi} - Z_W\right)\right] - 1\right\}$$

$$= \Delta\rho\left\{\frac{\Delta Z_H\sin\left[\frac{\pi(-u+Z_W)}{\Delta Z_H} + \alpha u\right]}{-2\pi + 2\alpha\Delta Z_H} + \frac{\Delta Z_H\sin\left[\frac{\pi(u-Z_W)}{\Delta Z_H} + \alpha u\right]}{2\pi + 2\alpha\Delta Z_H} - \frac{\sin(\alpha u)}{\alpha}\right\}\Bigg|_{Z_{CH_2}}^{Z_W}$$

$$= -\frac{\Delta\rho}{\alpha}[\sin(\alpha Z_W) - \sin(\alpha Z_{CH_2})]$$

$$+ \frac{\Delta\rho}{2}\left(\frac{1}{\alpha + \frac{\pi}{\Delta Z_H}} + \frac{1}{\alpha - \frac{\pi}{\Delta Z_H}}\right)[\sin(\alpha Z_W) + \sin(\alpha Z_{CH_2})]. \tag{A.39}$$

Because our X-ray scattering intensity was measured in a relative scale, an overall scaling factor was necessary for a non linear least square fitting procedure. This means that $\Delta\rho$ can be absorbed in the scaling factor. Doing so means that the values of ρ_{Hi} and ρ_M resulting from a fitting procedure are relative to $\Delta\rho$.

A.4 Correction Due to Refractive Index

q_z needs to be corrected for index of refraction [4].

Let θ' and λ' be the true scattering angle and wavelength within the sample. The wavelength by an energy analyzer, λ, and the scattering angle calculated from a position on a CCD detector, θ are apparent. The correction is not necessary in the horizontal direction. The Snell's law gives

$$n \cos \theta = n' \cos \theta' \tag{A.40}$$

$$n\lambda = n'\lambda'. \tag{A.41}$$

For low angle X-ray scattering, the momentum transfer along z direction is

$$q_z = \frac{4\pi \sin \theta'}{\lambda'} \tag{A.42}$$

$$= \frac{4\pi n'}{n\lambda} \sin \theta' \tag{A.43}$$

$$= \frac{4\pi n'}{n\lambda} \sqrt{1 - \cos^2 \theta'} \tag{A.44}$$

$$= \frac{4\pi n'}{n\lambda} \sqrt{1 - \left(\frac{n}{n'} \cos \theta\right)^2}. \tag{A.45}$$

The apparent scattering angle, θ, is directly related to the vertical pixel position, p_z, by

$$\theta = \frac{1}{2} \tan^{-1} \left(\frac{p_z}{S}\right), \tag{A.46}$$

where S is the sample-to-detector distance. The typical units of S and p_z are in mm. In our experimental setup, $n = 1$ and $n' = 0.9999978$ for lipids at $\lambda = 1.18\,\text{Å}$. $S = 359.7\,\text{mm}$.

Bibliography

1. A.B. Rodriguez-Navarro, Registering pole figures using an X-ray single-crystal diffractometer equipped with an area detector. J. Appl. Crystallogr. **40**, 631–634 (2007)
2. J.L. Baker, L.H. Jimison, S. Mannsfeld, S. Volkman, S. Yin, V. Subramanian, A. Salleo, A.P. Alivisatos, M.F. Toney, Quantification of thin film crystallographic orientation using X-ray diffraction with an area detector. Langmuir **26**(11), 9146–9151 (2010)
3. Y. Lyatskaya, Y.F. Liu, S. Tristram-Nagle, J. Katsaras, J.F. Nagle, Method for obtaining structure and interactions from oriented lipid bilayers. Phys. Rev. E **63**(1), 0119071–0119079 (2001)
4. Y. Liu, *New method to obtain strcuture of biomembranes using diffuse X-ray scattering: application to fluid phase DOPC lipid bilayers.* PhD thesis, Carnegie Mellon University, 2003

Printed in the United States
By Bookmasters